CONSERVATION AGRICULTURE
Approach and Application

CONSERVATION AGRICULTURE
Approach and Application

S K Acharya
Sreemoyee Bera
Cornea Saha
Prabhat Kumar
Monirul Haque
Riti Chatterjee
Anwesha Mandal

2023
Scholars World
A Division of
Astral International Pvt. Ltd.
New Delhi – 110 002

ISBN: 9789359191065 (HB)

Published by : **Scholars World**
A Division of
Astral International Pvt. Ltd.
– ISO 9001:2015 Certified Company –
4736/23, Ansari Road, Darya Ganj
New Delhi-110 002
Ph. 011-4354 9197, 8130496929
E-mail: info@astralint.com
Website: www.astralint.com

PREFACE

The aftermath of green revolution in Indian has generated, alongside food security, the sludge of ecological insecurity as well. The soil is being eroded at a faster rate than we could have estimated, the ground water contamination and its indiscriminate depletion are commencing isochronously with adverse effect on ecology and health. Our skewed attention to high yielding exotic genes has added a fillip to erosion of local plant races and diversity. So, this is the right time to refocus our attention from further depletion to conservation of our fragile resource base to a renewed elevation of resilience and balance. The ecosystem services are the life line for any civilization and , which has been passing through a harsh anthropogenic disruption. The dwindling of organic carbon level from soil layers and increasing energy prodigal behavior due to land fragmentation and exodus of input application only add to the misery. The decline of livestock count over the past couple of decades vindicates strongly the re-orchestration of crop-orchard-livestock-fish integration as a dynamic interplay into the very ecosystem services and functioning. If we think agriculture for ten years, the conventional approach can offer a trade off; but if we are to move forwards for century and beyond, we have to restore the ecosystem and redesign our production science and ethos in agriculture. A strong policy at national level to socialize and commercialize conservation agriculture has also been the imperatives; else it will perform as a programme without an institutionalization process with ecological voices and economics assurances across the holdings and farming. The book is a time felt pronunciation as to how and what we can do with conservation agriculture with much of empirical and hard evidences.

Prabhat Kumar

CONTENTS

THE AUTHORS

Prof. (Dr.) Sankar Kr Acharya, former Head, Dept. of Agricultural Extension and Director, Extension Education, Bidhan Chandra Krishi Viswavidyalaya, Mohanpur, WB, born on 6th October, 1960, started his career as Assistant Professor at BCKV in 1988 and has been in teaching, research and extension over 30 years. An erudite teacher as well as an elegant speaker, Prof. Acharya has also been the topper and awarded the certificate of merit in his M. Sc. (Ag) academic career in Agricultural Extension. He is internationally acclaimed for his unique research domain of Social Entropy and Energy Metabolism, Social Ecology and Environmental Sociology, Enterprise Ecology Framework, Farm Stewardships: The Transition from Conservation, Technology Socialization Process, and The Social Ecology of Livelihood, which have got tremendous policy as well as scholastic importance. Research Publication:221 in National and International Journals. Book Publication:101 books authored / co-authored, covering topics Participatory Extension Planning and Management; Value addition and quality management, Extension Strategy for Natural Resource Management; Horticulture Management with Quality Issues, Technology Socialization Process; Fish bio-diversity; Income andLivelihood; Enterprise Management and Entrepreneurship Building; Women's participation in Panchayat; Gender Issues in Floriculture; Technology gap in Floriculture; Enterprise Synergy in Agriculture; Disaster Management; Social Metabolism, Farm Energy Metabolism, Community Perception on Climate Change , Social Entropy and Chaos, Research areas of Publication: Community Forest Management and Role performance by EDC and FPC; Social forestry and community participation (paper published in Indian Forester); Membership of Distinction: Chancellor Nominee on selection committee for professors' positions; Member of Agriculture Commission; Marketing and Extension Sub Committee, Govt. of West Bengal; Expert member WWF projects in BTR (Buxa Tiger Reserve Project); Expert member; DFID project on Primary Education (DPEP); REIA, Visited Italy, France, Germany, China , Sri Lanka and Bangladesh. Delivered 18 Key note addresses in International Congress, Chaired 19 Sessions in International conferences, Editors of a score of national and international journals. He has so far been awarded with 13 best paper awards in National and International Conferences Organizing Secretary of two International Conferences, held in the years, 2016 & 2018, Jointly Organized by Department of Agricultural Extension, BCKV, WB and KrishiSanskriti, New Delhi. His Present RG score: 26.19He has also been honoured to be selected as the Convener of Panel (PE-32) entitled The hunger, poverty and silence, of IUAES, University of Manchester, UK, 2013. Co-PI of ICAR and World Bank funded project on Conservation Agriculture. So, far he has successfully guided 16 Ph D scholars from BCKV CU, KU, Vidyasagar University. He is the fellow, ISEE, IARI, New Delhi

Ms. Sreemoyee Bera, born on 24th May 1997 in South 24 Parganas district of West Bengal, obtained her B.Sc. Honours in Agriculture from College of Agriculture, Burdwan, Extended Campus of Bidhan Chandra Krishi Viswavidyalaya, Mohanpur, West Bengal in the year of 2019. She had obtained her M.Sc. (Agriculture) in Department of Agricultural Extension from Bidhan Chandra Krishi Viswavidyalaya, Mohanpur, West Bengal in the year of 2021. She completed her M.Sc. under the supervision of Prof. S K Acharya and her research area was on Managing, Monitoring and Analyzing of Agro-ecological Services: Conservation and Stewardship Approaches of the Farmers.

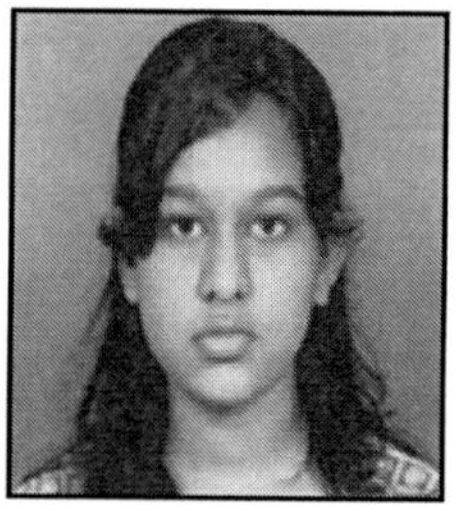

Ms.. Cornea Saha, born on 25th January 1997 in Alipurduar district of West Bengal, obtained her B.Sc. Honours in Agriculture from Uttar Banga Krishi Viswavidyalaya, Pundibari, West Bengal in the year 2019. Currently She is pursuing her M.Sc. (Agriculture) in department of Agricultural Extension from Bidhan Chandra Krishi Viswavidyalaya, Mohanpur, West Bengal under the supervision of Prof S K Acharya. She had worked under Centre of Advanced Agricultural Science and Technology (CAAST) on Conservation Agriculture, ICAR-NAHEP.

Dr.Prabhat Kumar, National Coordinator of Centre for Advanced Agricultural Science and Technology (CAAST) and Component 2 at ICAR-National Agricultural Higher Education Project, KAB-II, Pusa, New Delhi was born on 1st July 1973. He obtained his B. Sc. Degree in 1995 from C.S.A.Univ.of Ag. & Tech. Kanpur (U.P.) and M.Sc degree (Floriculture and Landscaping) in 1997 from Punjab Agril. University, Ludhiana, Punjab. He obtained his Ph. D degree in (Floriculture and Landscaping) in the year 2002 from IARI New Delhi. He joined G.B.P.U.A.&T. Pantnagar College of Agriculture, as Assistant Professor on 28th April 2003. He was selected to the post of Associate Professor on 31st March, 2012 at HNBGU, Srinagar, Garhwal (A Central University School of Agriculture). He was selected to the post of Senior Scientist at IARI, New Delhi on 23rd July, 2013. Dr.Prabhat Kumar is actively engaged in teaching of PG and Ph.D students and guiding PG students and Ph. D students at GBPUA&T, Pantnagar and IARI, New Delhi. To date, Dr.Prabhat Kumar has guided 10 M.Sc. and 01 Ph.D. students as Major Advisor in the discipline of Floriculture and Landscaping and presently guiding 02 Ph. D students. He has more than 18 years of experience in teaching, research and extension. He has handled more than 04 externally funded and 6 institute research projects and published 70 research papers in various national and international journals, 35 popular articles, 14book chapters, 11 technical bulletins and 06 books. He is associated in the development of 4 varieties namely; Pusa Deep, PusaBahar(Marigold)&PusaSinduri and Pusa Shanti(Gladiolus).

Dr Kumar has been honoured with four fellowships from prestigious scientific societies and awards in his carrier. Established Model Floriculture Centre (MFC) under Horticulture Technology Mission.

Mr Monirul Haque, born on 9th December 1995 in Alipurduar district of West Bengal, obtained his B.Sc. Honours in Agriculture from Palli Siksha Bhavana (Institute of Agriculture), Visva-Bharati, Sriniketan, West Bengal in the year of 2017. He had obtained his M.Sc. (Agriculture) in Department of Agricultural Extension from Bidhan Chandra Krishi Viswavidyalaya, Mohanpur, West Bengal in the year of 2019. He has qualified the National Eligibility Test for Assistant Professor conducted by UGC in his very first attempt in the year 2019. Currently, he is pursuing his Ph.D. in the Department of Agricultural Extension of the same University under the supervision of Prof S K Acharya. So far he has authored and co-authored 14 research articles in different national and international journals and co-authored 2 books. He has been awarded Maulana Azad National Fellowship for Minority Students for pursuing his research work.

Ms. Riti Chatterjee, born on 18th November in 1994, in Hooghly district of West Bengal, obtained her B.Sc. in Agricultural Sciences from Bidhan Chandra Krishi Viswavidyalaya, Mohanpur, West Bengal in 2016. She had obtained her M.Sc. (Ag.) in department of agricultural extension from the same university in 2018 as the First Class First in the university with OGPA 9.15, being the Gold-medallist. Her pursuits of hard work and diligence are reflected in her PG thesis, 'Occupational hazards and operating agro-ecosystem: the health and ecological dynamics of farm women', a unique empirical work under the exclusive supervision of Prof. S.K. Acharya and awarded best master's thesis award jointly by BHU and RASSA. So far she has attended 19 National and International Conferences, workshops and given oral and poster presentations with 5 best paper awards, and published 17 research papers on various national and international journals. She has already co-authored 7 books and 5 book chapters. She has qualified the National Eligibility Test for Assistant Professor conducted by UGC in her maiden attempt in May, 2017 and also the National Eligibility Test with ARS-Preliminary examination conducted by Agricultural Scientist Recruitment Board (ASRB) in May, 2018. Currently she is pursuing PhD in department of Agricultural Extension and working as Senior Research Fellow under Centre for Advanced Agricultural Science and Technology (CAAST) on Conservation Agriculture project under ICAR-NAHEP-BCKV. In the months of January to June 2020 she worked as short-term Visiting Scholar at The Ohio State University, Ohio, USA and there studied on Social-ecology of Conservation Agriculture at stakeholder level.

Ms. Anwesha Mandal was born on 25th October, 1993 at Kalyani, West Bengal. She did her graduation in Agriculture from Palli Siksha Bhavana, Visva Bharati. In 2016, she joined in the Department of Agricultural Communication at G. B. Pant University of Agriculture and Technology (GBPUAT), Pantnagar, in Uttarakhand for completing her Post Graduation with ICAR Junior Research Fellowship (NTS). Presently, she has submitted her Ph.D. thesis at Bidhan Chandra KrishiViswavidyalaya (BCKV), Mohanpur, Nadia, in the Department of Agricultural Extension as an SRF in the CAAST project on Conservation Agriculture under the guidance of Prof. S.K. Acharya. She had cleared both ICAR-National Eligibility Test (NET) in Agricultural Extension and UGC- National Eligibility Test (NET) in Adult Education with JRF scholarship in her maiden attempt. She has worked on Soil Health aspect in her M.Sc. Thesis. Besides, she also cleared other national level examinations held by ICAR and other deemed universities of Agriculture. She participated in national and international seminars and also received best paper presentation award from ISEE National Conference, 2018 in Kolkata. She has already been co-authors of three books in the domain of Extension Science.

PART – I
Managing, Monitoring and Analyzing Agro-ecological Services: Conservation and Stewardship Approaches of the Farmers

Synopsis

New alluvial zone of Eastern India is characterized with high intensity agriculture which in turn is both coercive and depletive by nature. Almost 120 blocks in West Bengal are under the clutch of arsenic contamination which is enjoying a pervasive passage through the food chains to kill both human being and livestock. High intensity mechanization, as it is becoming exponentially popular due to lack of agricultural labourers associated with unaffordable wage, is ushering soil erosions and declining organic carbon level. The sharp decline of livestock population can be correlated with the lowering of organic carbon status. The social ecology of this zone is predominant by farmers who are mostly owner of small, marginal and fragmented holdings. This has further made the tiny holdings energy prodigal and economically non-manageable. The present study was conducted on 50 farmers, small and marginal farmers in Dalilpur and Kastodanga village locales. A structured schedule has been administered to generate responses against a shade of queries. Ecological resilience has been estimated by using a set of dummy variables by calculating the slope and pattern of changes of crop residue admixture, yield topography, livestock count, application of organic manure and change in cropping intensity. The study further elicits that the variables like climate resilient technology, biodiversity conservation perception, soil stewardship perception and water stewardship perception, performance and product monitoring have contributed substantially to the regime and level of ecological resilience as have characterized the dynamics of agriculture. ANN (Artificial neural network) and Path analyses have been applied to estimate the role, function and contribution of respective variables to the criterion character i.e. ecological resilience in conservation agriculture. The study offers huge micro sociological policy in dealing with diverse issues of conservation agriculture towards attaining sustainability and productivity through socializing conservation agriculture across the small and marginal land holders in India.

Key words: *Conservation Agriculture, Ecological Resilience, Stewardship perception, Monitoring and tracking, Artificial Neural Network*

1

Introduction

Since the beginning, the ecosystem has been providing human beings with many services in the form of food, forage, bio-energy and pharmaceuticals. Over the centuries we too often took these benefits for granted and underestimated the impacts of our activities on nature. Current global change patterns are threatening the supply of agrarian ecosystem services on which human well-being depends. Ecosystem services are all the benefits to humans from the ecosystems. Ecosystem services were classified into the following categories of services by the Millennium Ecosystem Assessment (2005).

1. Provisioning Services which includes the products/raw materials or energy outputs like food, water, medicines and other resources from ecosystems.
2. Regulating Services are ecosystem functions that make the system deterrent of impurities, pollutants and regulate inhibitors to ecosystem resilience.
3. Supporting Services form the basis for other services. These services provide habitat for different life forms, retain biodiversity, nutrient cycling, and other services for supporting life on the earth.
4. Cultural Services includes tourism; provides recreational, aesthetic, cultural and spiritual services etc.

The aftermath of green revolution has not only contributed to a 'quantum jump' of agricultural production and productivity, from 55 MT (1955) to 120 MT (1970), but also to a plethora of deleterious impacts on the natural resources and ecological health. Intensive farming practices has destroyed huge regions of natural habitat and caused an untold loss of ecosystem services, and it is responsible for about 30% of greenhouse gas emissions. There has been a fall out in the ecological balances due to the injudicious use of agrochemicals associated with farm mechanization which led to soil erosion and compaction. Every year, around 15–18 tonnes of top soils are eroded off to add to the impoverishment of soil regime. On the other side, the energy prodigal nature of agriculture along with over exploitation of ground water have added to the misery of our key production ecology. The State of Food Security and Nutrition in the World report estimated that almost 690 million people globally were undernourished in 2019. Yet to feed the global population expected by 2050, more than 1 billion hectares of wild land need to be converted to farmland if current farming approaches continue to remain used. Therefore, to feed the world without further damaging the ecosystem, conventional agricultural practices must be transformed.

With this perspectives, conservation agriculture and stewardship approaches for managing and monitoring of ecological services is both prominent and pertinent. Conservation agriculture is mainly a concept for resource-saving agricultural crop production to achieve acceptable profits along with high and sustainable production levels subsequently conserving the environment" (FAO 2007). Conservation agriculture (CA) is a set of technologies, including minimum soil disturbance, permanent soil cover, diversified crop rotations, and integrated weed management (Reicosky and Saxton 2007; Hobbs etal. 2008; Friedrich et al. 2012), aimed at reducing and/or reverting many negative effects of conventional farming practices such as soil erosion (Putte etal. 2010), soil organic matter (SOM) decline, water loss, soil physical degradation, and fuel use (Baker etal. 2002; FAO 2008). Now the two aspects conservation and stewardship need to be operationally discriminated. Conservation is centering around maintenance to protect the count of resource and diversity whereas stewardship focuses on mentoring with a purpose to socio-ecological resilience. Ecological resilience, also called ecological robustness is the ability of an ecosystem to maintain its normal patterns of nutrient cycling and biomass production after being subjected to damage caused by an ecological disturbance. Accustoming and upgrading conservation practices with stewardship function are adaptable and serializable functions to attain a sustainable growth and development. This is the process where in forest is protected from its biodiversity depletion and at the same time the part of the forest is opened up for ecotourism. Thus, conservation elements are extended to stewardship functions by generating both income and livelihood. Stewardship function welcomes ecological resilience by incubating and internalizing ecological deliverables to support both human and non-human life forms. Ecological functions are basically the productive performance of ecological deliverables for the surrounding social-ecosystem to make it an agile, dynamic and ever sprouting life support system.

The history of evolving agriculture and ecology has been co-genital so far. The trend and direction of change in agriculture gets performing in resonance with ecological functions. The health of soil and its productive functions are correlated with organic carbon status, which merits a reciprocation from crop residues mixing with soil and application of organic manure. The interactive functions between livestock resources and soil fertility status, biodiversity and agricultural sustainability, social ecology and issues of income-livelihood have got a telling impact in sustaining productivity across the time and space. Farm energy metabolism speaks to the very eco-dynamics of the entire agricultural production system and production behaviour at large. So our farm ecology has to go energy efficient for retaining the ecological function effectively over a protractile period. So economy,ecology, energy have come out as to be the most important determinants for earning integrated sustainability and regenerating productivity along the undulations of agro-ecological evolution.

Rationale

Conventional agriculture has turned both genetically fatigued and operationally tired as reflected through plateauing of agricultural productivity and quality as well, so we need conservation agriculture so as to-

1. Restore soil organic carbon, checking soil erosion, reducing no. and frequency of tillages, improving crop biodiversity and assuring sustainable livelihood.
2. Participatory, monitoring, tracking, auditing and operating are essential to rationalize tillage intensity, water and energy exuberance in farming.

Our ecological functions and services as operating through agro-ecosystem resilience should be enhanced, restored and synchronized.

3. The study thus justifies the objective sets here in order to delineate the pathway for comprehensive, socialization of conservation agriculture by creating robust platform in both socio-ecological scalability and non-scaling as well.

Objectives

1. Delineation of the general status of managing, monitoring and analyzing agro-ecological services in conservation and stewardship.
2. Identification and customization of variables related to monitoring, managing and analyzing (impacting on agro-ecological services) in terms of conservation and stewardship.
3. To estimate the interaction and impact of different system variables on the agro-ecological services for conservation and stewardship.
4. Participatory analysis and interpretation of monitoring, managing and analyzing agro-ecological services towards achieving conservation and stewardship goal.
5. Generation of micro sociological and ecological policy for attaining desired level of conservation and stewardship goal.

Limitations

1. More number of relevant variable could have been included.
2. Number of respondents has been restricted by Covid-19 situation, else it could have been more at least 100.
3. Complete random sampling of respondent has not possible due to Covid-19, hence, snowballing approach was followed.
4. In certain cases, when pandemic was severe, had to depend on telephonic conversation for data collection.
5. Sensitivity analysis could have added a better implication along with artificial neural network.

2

Theoretical Orientation

This chapter deals with some important terms and their concepts related to the study which are essential to convey the ideas of the study effectively and also to provide a theoretical base for the empirical investigation and guidance for the selection of relevant predictor variables.

2.1. Conceptual Framework

A conceptual framework is an analytical tool that is used to get a comprehensive understanding of the study. It embodies the specific direction by which the research will have to be undertaken. It is generally constructed to provide the researcher the element of attainable and researchable problems and ideas supported by methodology, design, prospective analysis and empirical conclusion. The conceptual framework is often called as the research paradigm.

The subject matter of the present research is **Managing, Monitoring and Analyzing Agro-ecological Services: Conservation and Stewardship Approaches of the Farmers.** It is expected that the terminologies and concepts used to frame up the topic need to undergo brief explanation for the better understanding of its meaning.

Managing: the process of dealing with or withstanding (something) to accomplish a definite goal within a definite time frame.

Monitoring: continuous process of collecting and analyzing information for observing and checking the progress or quality of (something) over a period of time; to keep track of or maintaining regular surveillance over (something).

Analyzing: to examine (something) methodically and in detail, typically in order to explain and interpret it.

Ecological Services: Ecosystem services or ecological services are mainly, benefits to humans from the ecosystems. Ecosystem services were classified into the following four categories of services by the Millennium Ecosystem Assessment (2005)-

i. Provisioning Services

ii. Regulating Services

iii. Cultural Services

iv. Supporting Services

Conservation: It is centering around maintenance to protect the count of resource and diversity.

Stewardship: It is a strategy to respond to and shape social–ecological systems under conditions of uncertainty and change to sustain the supply and opportunities for use of ecosystem services to support human well-being.

2.2. A Framework for Understanding Change: Conservation and Stewardship

The world is undergoing unprecedented changes in many of the factors that determine both its fundamental properties and their influence on society. Throughout human history, people have interacted with and shaped ecosystems for social and economic development (Turner et al. 1990, Redman 1999, Jackson 2001, Diamond 2005). During the last 50 years, however, human activities have changed ecosystems more rapidly and extensively than at any comparable period of human history (Steffen et al. 2004, Foley et al. 2005, MEA 2005d; Plate 1). Earth's climate, for example, is now warmer than at any time in the last 500 (and probably the last 1,300) years (IPCC 2007a), in part because of atmospheric accumulation of carbon dioxide (CO2) released by the burning of fossil fuels (Fig. 1.1). Agricultural development largely accounts for the accumulation of other trace gases that contribute to climate warming. As human population increases, in part due to improved disease prevention, the increased demand for food and natural resources has led to an expansion of agriculture, forestry, and other human activities, causing large-scale land-cover change and loss of habitats and biological diversity. About half of the world's population now live in cities, and depend on connections with rural areas worldwide for food, water, and waste processing. In addition, increased human mobility is spreading plants, animals, diseases, industrial products, and cultural perspectives more rapidly than ever before. This increase in global mobility, coupled with increased connectivity through global markets and new forms of communication, links the world's economies and cultures, so decisions in one place often have international consequences.

This **globalization** of economy, culture, and ecology is important because it modifies the **life-support system** of the planet (Odum 1989), i.e., the capacity of the planet to meet the needs of all organisms, including people. The dramatic increase in the extinction rate of species (100- to 1,000-fold in the last two centuries) indicates that global changes have been catastrophic for many species, although some species, especially invasive species and some disease organisms, have benefited and expanded their ranges. Human society has both benefited and suffered from global changes, with increased food production, increased income and living standards (in parts of the world), improved treatment of many diseases, and longer life expectancy being offset by deterioration in **ecosystem services**, the benefits that society receives from ecosystems. More than half of the ecosystem services on which society depends for survival and a good life have been degraded—not deliberately, but inadvertently as people seek to meet their material desires and needs (MEA 2005d). Change creates both challenges and opportunities.

The unique feature of the changes described above is that they are **directional**. In other words, they show a persistent trend over time (Fig. 1.1). Many of these trends have become more pronounced since the mid-twentieth century and will probably continue or accelerate in the coming decades, even if society takes concerted actions to reduce some rates of change. This situation creates a dilemma in planning for the future because we cannot assume that the future world will behave as we have known it in the past or that our past experience provides an adequate basis to plan for the future. This issue is especially acute for sustainable management of natural resources. It is no longer possible to manage systems so they will remain the same as in the recent past, which has traditionally been the reference point for resource managers and conservationists. We must adopt a more flexible approach to managing resources—management to sustain the functional properties of systems that are important to society under conditions where the system itself is constantly changing. Managing resources to foster resilience—to respond to and shape change in ways that both sustain and develop the same fundamental function, structure, identity, and feedbacks—seems crucial to the future of humanity and the Earth System. Resilience-based ecosystem stewardship is a fundamental shift from steady-state resource management, which attempted to reduce variability and prevent change, rather than to respond to and shape change in ways that benefit society (Table 1.1). We emphasize resilience, a concept that embraces change as a basic feature of the way the world works and develops, and therefore is especially appropriate at times when changes are a prominent feature of the system. We address ecosystems that provide a suite of ecosystem services rather than a single resource such as fish or trees. We focus on stewardship, which recognizes managers as an integral component of the system that they manage. Stewardship also implies a sense of responsibility for the state of the system of which we are a part (Leopold 1949). The challenge is to anticipate change and shape it for sustainability in a manner that does not lead to loss of future options (Folke et al. 2003). Ecosystem stewardship recognizes that society's use of resources must be compatible with the capacity of ecosystems to provide services, which, in turn, is constrained by the life-support system of the planet (Fig. 2.2).

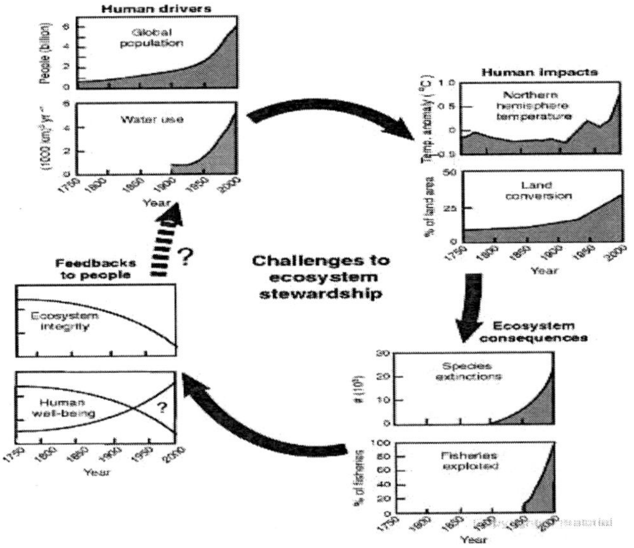

Figure 2.1: Challenges to ecosystem stewardship. Changes in human population and resource consumption alter climate and land cover, which have important ecosystem consequences such as species extinctions and over-exploitation of fisheries. These changes reduce ecosystem integrity and have regionally variable effects on human well-being, which feeds back to further changes in human drivers. Panel inserts redrawn from Steffen et al. (2004).

Table 2.1: Contrasts between steady state resource management, ecosystem management, and resilience based ecosystem stewardship

Steady state resource management	Ecosystem management	Resilience based ecosystem Stewardship
Reference state: historic condition	Historic condition	Trajectory of change
Manage for a single resource or species	Manage for multiple ecosystem services	Manage for fundamental social–ecological properties
Single equilibrium state whose properties can be sustained	Multiple potential states	Multiple potential
Reduce variability	Accept historical range of variability	Foster variability and diversity
Prevent natural disturbances	Accept natural disturbances	Foster disturbances that sustain social–ecological properties
People use ecosystems	People are part of the social–ecological system	People have responsibility to sustain future options
Managers define the primary use of the managed system	Multiple stakeholders work with managers to define goals	Multiple stakeholders work with managers to define goals

Steady state resource management	Ecosystem management	Resilience based ecosystem Stewardship
Maximize sustained yield and economic efficiency	Manage for multiple uses despite reduced efficiency	Maximize flexibility of future options
Management structure protects current management goals	Management goals respond to changing human values	Management responds to and shapes human values

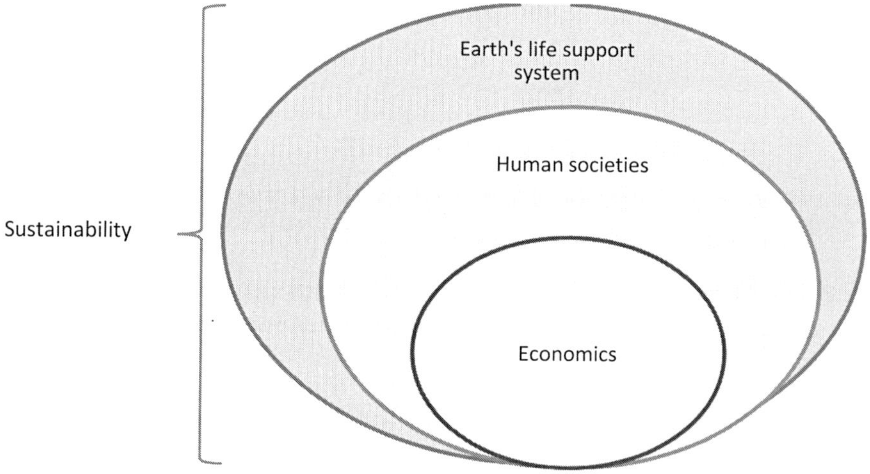

Figure 2.2: Social–ecological sustainability requires that society's economy and other human activities not exceed the capacity of ecosystems to provide services, which, in turn, is constrained by the planet's life support system. Redrawn from Fischer et al. (2007).

This figure delineated above presents a framework for understanding and managing resources in a world where persistent directional changes are becoming more pronounced.

2.3. The 10 Elements of Agro-ecology

The 10 Elements of Agroecology resulted from a multi-stakeholder process intended to generate a system re-design framework to be optimized and adapted to local contexts. The 10 Elements of Agroecology framework was originally developed in 2015. Prominent themes from presentations delivered during the **First International Symposium on Agroecology for Food Security and Nutrition** provided an initial coherent structure: recycling, efficiency, diversity, resilience and synergies as central ecological features of agroecology. Nevertheless, calls in regional meetings for reinforcing social and political aspects of agroecology were also strong. Thus, these aspects that emerged from regional consultations were clustered under an additional five elements: co-creation of knowledge; human and social values; culture and food traditions; responsible governance; and circular and solidarity economy. Following the refinement of element names, content, and the development of a consistent storyline highlighting the interconnected nature of agroecology and its 10 Elements, the framework was finalized after several rounds of review by international and FAO experts.

The 10 Elements of Agroecology framework was launched at the **Second FAO International Symposium on Agroecology** held in April 2018 and continues to evolve. In October 2018, the 10 Elements of Agroecology were supported by the FAO Committee on Agriculture (COAG) at its 26th Session as a guide to one of the ways to promote sustainable agriculture and food systems. Following the review, revision, and clearance process through FAO's governing bodies, the 10 Elements of Agroecologywere approvedby the 197 Members of FAO to guide FAO's vision on agroecology by the 163 session of the Council on 2-6 December 2019.

1. **Diversity:** diversification is key to agroecological transitions to ensure food security and nutrition while conserving, protecting and enhancing natural resources.
2. **Co-creation and sharing of knowledge:** agricultural innovations respond better to local challenges when they are co-created through participatory processes.
3. **Synergies:** building synergies enhances key functions across food systems, supporting production and multiple ecosystem services.
4. **Efficiency:** innovative agroecological practices produce more using less external resources.
5. **Recycling:** more recycling means agricultural production with lower economic and environmental costs.
6. **Resilience:** enhanced resilience of people, communities and ecosystems is key to sustainable food and agricultural systems.
7. **Human and social values:** protecting and improving rural livelihoods, equity and social well-being is essential for sustainable food and agricultural systems.
8. **Culture and food traditions:** by supporting healthy, diversified and culturally appropriate diets, agro ecology contributes to food security and nutrition while maintaining the health of ecosystems.
9. **Responsible governance:** sustainable food and agriculture requires responsible and effective governance mechanisms at different scales – from local to national to global.
10. **Circular and solidarity economy:** circular and solidarity economies that reconnect producers and consumers provide innovative solutions for living within our planetary boundaries while ensuring the social foundation for inclusive and sustainable development.

2.4. Ecological Resilience and Conservation Agriculture

The food choice that precipitated the COVID-19 crisis was predicted more than a decade ago,1 and the disruption induced by the virus fits a model of **adaptive cycle** dynamics developed nearly 50 years ago (Holling, 1973). Adaptive cycles of rapid growth, maturity, release, and reorganization are ubiquitous in social-ecological systems. Ecological resilience refers to the degree of disturbance a system can buffer before entering the collapse/release and reorganization phases. Systems can induce collapse and reorganization that affects the resilience of their subsystems as well as other systems (Sundstrom & Allen, 2019). In recent years the concept of ecological resilience has been applied most vigorously to food systems and climate change. Agriculture is affected by climate change perhaps more than any other sector of our society (National Sustainable Agriculture Coalition [NSAC], 2019). One recent policy response is the Agriculture Resilience Act (2020), which seeks to mitigate climate change by sequestering carbon and reducing other greenhouse gases through processes that enable food systems to cope with climate change by increasing soil health and, thereby, yields and profits.

The concept or metaphor of the **adaptive cycle was developed by Holling (1986)** to describe the different phases of system behavior in (managed) ecosystems. These systems show a tendency to repeat characteristic behavioral phases e.g. succession states. The cycle was first used to describe temperate ecosystems (e.g. boreal coniferous forests, grassland). Later it was also adapted to ecosystems in other regions of the world and even to human organizations (e.g. bureaucratic) and economics. In a traditional ecology view of the succession processes two phases are emphasized: The first is the rapid colonization of disturbed areas (exploitative phase). The second is the conservation, the slow accumulation and storage of energy and material in established (Climax) ecosystems (conservation phase). The organisms engaged in the exploitive phase are termed r-strategists the ones predominant in the conservation stage K-strategists. The r types are usually so-called pioneer species, characterized by exponential population growth – often leading to a final collapse – and excellent dispersal strategies. The K – strategists are seen to maintain their population close to some maximum sustainable population. Generally, they have less offspring (but invest more into them) and are more specialized then the r-strategists. Holling now adds two additional functions: (1) the release or "creative destruction" (a term borrowed from Schumpeter (1950)) (2) the reorganization phase. The release phase is termed O the reorganization phase a. In a two-dimensional view the adaptive cycle has two properties: The potential and the connectedness. In the K-phase of a system the accumulation (potential) of biomass and nutrients is high and tightly bound (over connected in system terms). This leads to an increasingly fragile system. Then some sudden events release the biomass and nutrients and lead to the release or O phase). Such events can be: forest fires, drought, insect pest, or intense pulses of grazing. The ensuing phase is the reorganization or a phase. In this phase the nutrient loss is minimized and the nutrient availability reorganized. The pioneer species appear and utilize accumulated and emerging energy and nutrient sources.

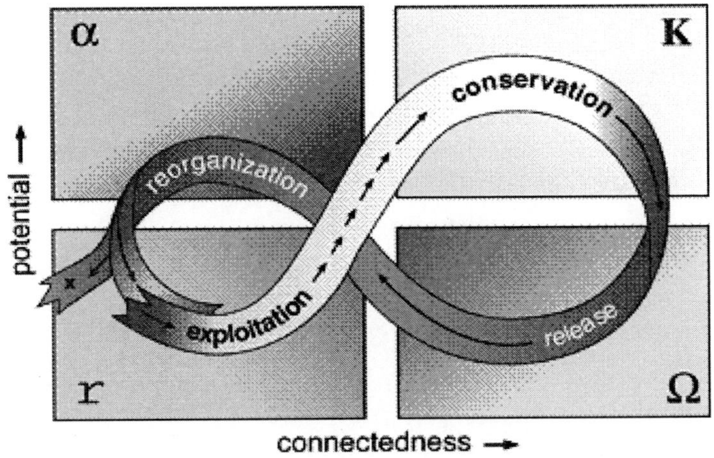

Figure 2.3: The "lazy eight" is a stylized illustration of the four ecosystem functions (r, O, K a), and the flow of events among them. The length of the arrows indicates the speed of change in the different phases (short arrows = fast; long arrows = slow). The x-axis shows the degree of connectedness the y-axis the potential of the accumulated resources. (In Gunderson et al, 2002)

The resilience of socio-ecological systems is described by Carpenter et al (2002) as the amount of disturbance a system can absorb and still remain within the same state. The degree to which the system is capable of self-organization (versus lack of organization, or organization forced by external factors) and the degree to which the system can build and increase the capacity for learning and adaptation. The greater the resilience of the system the easier it will absorb shocks and be capable to adapt to changes. The adaptive capacity of a society is constraint by its institutions and by the environment. Social-ecological resilience is determined by the livelihood security of an individual or groups. This security involves the access and entitlement to natural resources (Berkes et al 2003). The resilience of social system is also influenced by the interaction of different levels of hierarchies.

Resilience depends on:

1. Adaptive capacity
2. Biophysical and social legacies that contribute to diversity and provide proven pathways for rebuilding
3. The capacity of people to plan for the long term within the context of uncertainty and change
4. A balance between stabilizing feedbacks that buffer the system against stresses and disturbance and innovation that creates opportunities for change

5. The capacity to adjust governance structures to meet changing needs (Holling and Gunderson 2002, Folke et al.2003, Walker et al.2006)

The analytical use of resilience began in ecology, mainly within ecological stability theory and Holling's (1973) definition of resilience as a measure of the persistence of systems and of their ability to absorb change and disturbance and still maintain the same relationships between populations or state variables. Within ecology there has been a high level of conceptual confusion regarding resilience and related terms (Grimm and Wissel 1997).

One primary difference between resilience and robustness is the strong connection between resilience and other concepts such as attractors, states, steady states, stability, domains of attraction, basins of attraction, regimes, and equilibria, none of which are referenced frequently by literature on system robustness.

Conservation agriculture is mainly a concept for resource-saving agricultural crop production to achieve acceptable profits along with high and sustainable production levels subsequently conserving the environment" (FAO 2007). So, it has evolved as an alternative approach to conventional agricultural systems in which tillage is practised to varying levels to control weeds, pests, nutrient incorporation and to check soil compaction problems where conservation agriculture helps to maintain, to the extent possible, a year-round soil cover by residue retention from previous crops or incorporation from another field/crop and/or a cover crop to improve the soil quality and minimal soil disturbance by no or reduced tillage practices and through crop diversification and/or crop rotations to restrict the pest problems (Dillaha et al.).

Enhanced resilience of people, communities and ecosystems is key to sustainable food and agricultural systems. Diversified agroecological systems are more resilient – they have a greater capacity to recover from disturbances including extreme weather events such as drought, floods or hurricanes, and to resist pest and disease attack. Following Hurricane Mitch in Central America in 1998, biodiverse farms including agroforestry, contour farming and cover cropping retained 20–40 percent more topsoil, suffered less erosion and experienced lower economic losses than neighbouring farms practicing conventional monocultures.

By maintaining a functional balance, agroecological systems are better able to resist pest and disease attack. Agroecological practices recover the biological complexity of agricultural systems and promote the necessary community of interacting organisms to self-regulate pest outbreaks. On a landscape scale, diversified agricultural landscapes have a greater potential to contribute to pest and disease control functions.

Agroecological approaches can equally enhance socio-economic resilience. Through diversification and integration, producers reduce their vulnerability should a single crop, livestock species or other commodity fail. By reducing dependence on external inputs, agroecology can reduce producers' vulnerability to economic risk. Enhancing ecological and socio-economic resilience go hand-in-hand – after all, humans are an integral part of ecosystems.

2.5. Ecosystem Services and Conservation Agriculture

Conservation agriculture (CA) has emerged as a promising technology for efficient rational use of available resources and sustained productivity in the long run. By saving inputs, reducing energy usage and greenhouse gases emissions, CA-based management practices are quite viable for bringing sustenance in agricultural crop production. The CA system can provide multiple ecosystem services such as provisioning, regulating and supporting services. The regulating services include improving carbon status, and physical, chemical and biological properties of soil, which further lead to provisioning services in terms of sustained crop and water productivity. Increased soil carbon sequestration improves supporting services, namely, soil aggregation that increases available soil moisture and can be helpful for better plant growth and development. It also improves soil biodiversity both above-and below-ground. Here we focus on the potential ecosystem service benefits accrued from CA. Conservation agriculture in the long run can be a strategy for sustainable crop intensification and a climate resilient crop management system.

CA is described by three interlinked principles, along with other good agricultural practices, namely: (i) continuous no or minimal mechanical soil disturbance (implemented by the practice of no-till seeding or broadcasting of crop seeds, and direct placing of planting material into untilled soil; and causing minimum soil disturbance from any cultural operation, harvest operation or farm traffic); (ii) maintenance of a permanent biomass soil mulch cover on the ground surface (implemented by retaining crop biomass, root stocks and stubbles and cover crops and other sources of ex situ biomass); and (iii) diversification of crop species; implemented by adopting a cropping system with crops in rotations, and/or sequences and/or associations involving annuals and perennial crops, including a balanced mix of legume and non-legume crops (Kassam et al. 2018). Controlled traffic that lessens soil compaction could be another principle being pondered upon in the recent years. CA is a promising technology for efficient rational use of available resources and sustained productivity in the long-run (Das et al. 2014, Nath et al. 2017, Oyeogbe et al. 2018).

Globally, CA is being practiced on about 125 M ha. The major CA practicing countries are USA (26.5 M ha), Brazil (25.5 M ha), Argentina (25.5 M ha), Canada (13.5 M ha) and Australia (17.0 M ha). CA is being adopted in varying degrees in south – east Asia i.e., Japan, Malaysia, Indonesia, Korea, Philippines, Taiwan, Srilanka and Thailand. No tillage based CA systems are gaining increased attention in Africa especially of southern and eastern Africa.

In India, efforts to adopt and promote conservation agriculture technologies have been underway for nearly a decade but it is only in the last 8 – 10 years that the technologies are finding rapid acceptance by farmers. Efforts to develop and spread conservation agriculture have been made through the combined efforts of several State Agricultural Universities, ICAR institutes and the Rice-Wheat Consortium for the Indo-Gangetic Plains. The spread of technologies is taking place in India in the irrigated regions in the Indo-Gangetic plains where rice-wheat cropping systems dominate.

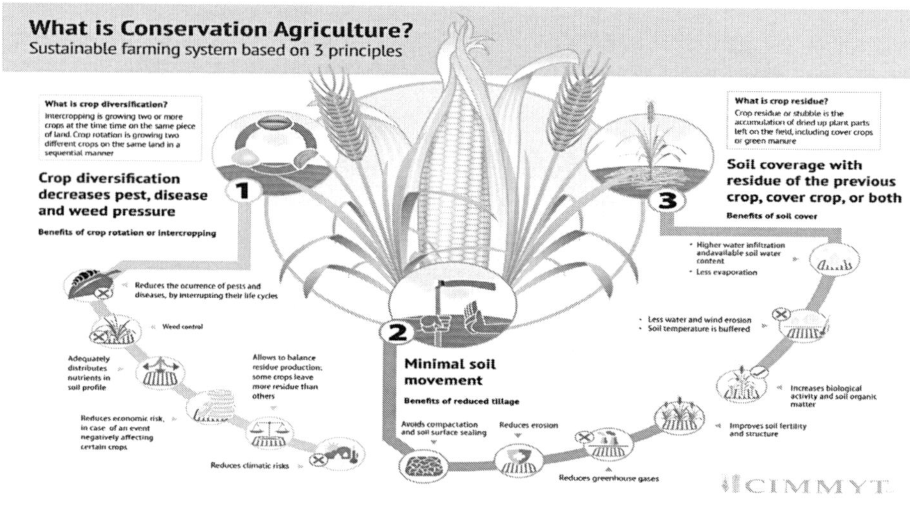

Figure 2.4: Conservation Agriculture and its principles

There are many potential ecosystem services associated with the benefits of conservation agriculture. Ecosystem services are natural processes through which the environment produces natural resources that humans and other living species require for life (Dillaha et al. 2010). These include the benefits that human derive directly or indirectly from ecosystem functions (Costanza et al. 1997). Millennium Ecosystem Assessment (2005) has grouped ecosystem services into four broad categories: provisioning, supporting, regulating and cultural services with specific services mentioned in each category. These are presented in Fig 3.4 with some modifications/additions of specific services. Although CA was originally introduced to regulate wind and water erosion (Baveye et al. 2011), is now considered to deliver multiple ecosystem services (Oyeogbe et al. 2017, 2018).

2.5.1. Potential ecosystem services from conservation agriculture

Several (provisioning, supporting, regulating and cultural) ecosystem services are enlisted by Millennium Ecosystem Assessment (2005). They all are not influenced by CA practices. The ecosystem services, which are positively influenced by CA are discussed below.

- **Crop/food production:** Conservation tillage, the most important aspect of CA is thought to take care of the soil health, plant growth and environment (Bhattacharyya et al. 2013, 2015). Conservation tillage has beneficial effects on crop root growth, water and nutrient use efficiencies and ultimately the agronomic yield (Das et al. 2018). It has been observed that under drought and high temperature conditions, no-till wheat is more resilient than conventional wheat crop. Das et al. (2018) reported that in CA-based maize-wheat cropping system, maize yields due to permanent broad bed with residue and permanent narrow bed with residue were 28 and 15% higher than that in CT plots (Table 1). Singh et al. (2016) reported that CA-based management practices such as dry direct-seeded rice (DSR), zero tillage and residue retention may hold potential to increase yields, reduce costs and increase farmers' profits in rice-maize system (RMS). They observed that root mass density was 6 to 49% greater in rice and 21 to 53% in maize under zero-tilled DSR followed by zero-tilled maize with residue retention from both crops compared to conventional RMS in different soil layers to 60 cm depth. Several other studies revealed that zero tillage wheat seeding technology can play an important role in saving inputs, turnaround time, reducing energy usage and environmental pollution, enhancing productivity and increasing farmers' income (Kumar et al. 2013; Das et al. 2014, 2016, 2018).

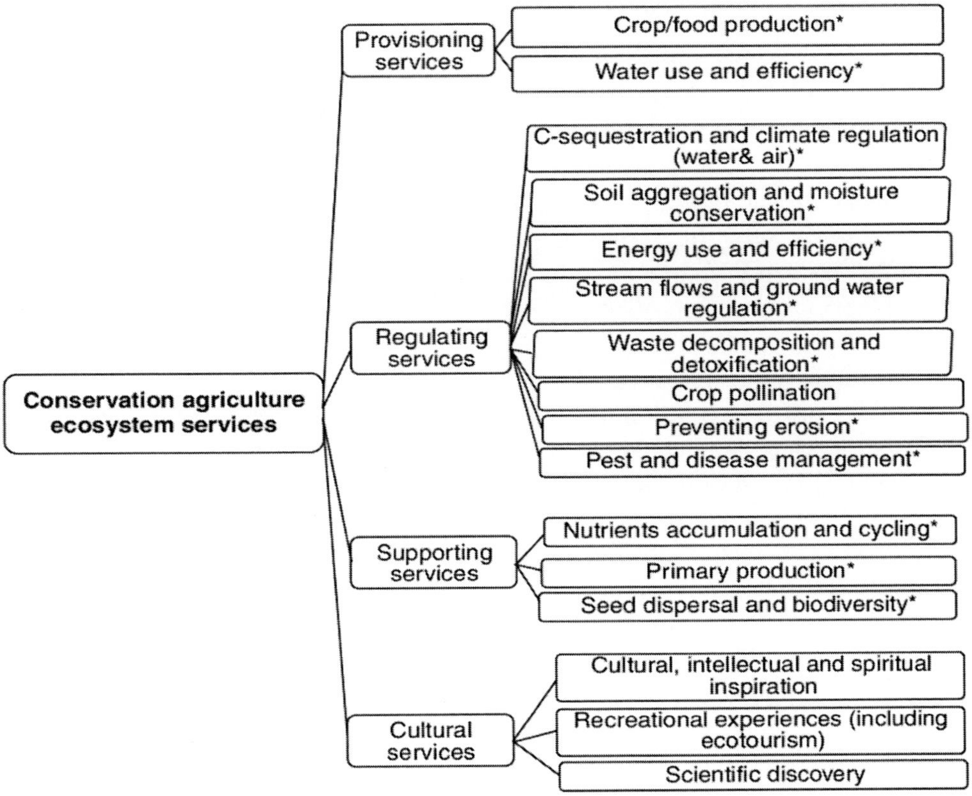

Figure 2.5: Different ecosystem services provided by conservation agriculture (*services are positively influenced by CA) (Source: Millennium Ecosystem Assessment 2005, Dillaha et al. 2010, modified by Authors)

Table 2.2: Crop productivity in conservation agriculture (CA) and conventional tillage (CT) treatments

Conservation/ zero tillage	Study duration (Years)	Crop	Increase in crop productivity over CT (%)	Reference
Permanent broad bed with residue	3	Cotton (Cotton wheat)	51.0 (Cotton)	Das *et al.* (2014)
Permanent broad bed with residue	3	Maize (Maize wheat)	28.0 (Maize)	Das *et al.* (2018)
Zero-till flat bed	4	Pigeon pea	5.4	Sepat *et al.* (2015)
		Wheat	2.5	
ZT DSR – ZTW+R	7	Rice-wheat	1.3 Mg/ha/ year	Jat *et al.* (2014)
Zero-till wheat (ZTW)	3	Wheat	9.7	Kumar *et al.* (2013)
No till with residue cover	15	Wheat	19.1	Li et al. (2007

- **Carbon sequestration and climate regulation:** One of the most positive aspects of conservation agriculture is its contribution to increase soil carbon compared with tillage-based conventional agriculture systems. Lal (2004) in his studies of the benefits of increasing soil carbon on its potential for carbon sequestration for climate regulation found that the carbon sink capacity of the world's agricultural and degraded soils is 50 percent to 66 percent of the global historic carbon loss of 42 to 78 gigatons of carbon and that improved land management practices on the world's agricultural and degraded soils could sequester 50 percent to 66 percent of the historic soil carbon loss. This is equivalent to 0.4 to 1.2 Gt C/year, or 5 percent to 15 percent of global carbon emissions. Lal also stated that the rate of increase in the Soil Organic Carbon (SOC) stock, through recommended management practices, follows a sigmoid curve, attains the maximum 5 to 20 years after adoption of recommended management practices, and continues until SOC attains a final equilibrium. Hillel and Rosenzweig (2009) also studied that the conversion to no-till farming increased soil organic carbon by rates varying from 0.1 to 0.7 Mg/ha-yr and, like Lal, indicated that such positive increments cannot be expected to continue indefinitely as well managed carbon depleted soils will tend to approach their natural equilibrium (or C saturation) state within a few decades. Lal (2004) reported similar rates of soil organic carbon sequestration in agricultural and restored ecosystems depending on soil texture, profile characteristics, and climate, which ranged from 0 to 0.15 Mg/ha-year in dry and warm regions, and 0.1 to 1.0 Mg/ha-yr in humid and cool climates.

- **Water use and efficiency:** The provision of sufficient quantities of clean water is an essential ecological service to agro-ecosystems, and agriculture accounts for about 70% of global water use (WHO 2003). Water availability in agro ecosystems depends not only on infiltration and flow, but also on soil moisture storage, another type of ecosystem service (Power 2010). About 80% of agricultural water use comes from rainfall and stored as soil moisture (~green water). The supply of surface water and groundwater (~blue water) inputs to agriculture through irrigation are indispensable in some parts of the world. With climate change, increased variability of rainfall is predicted to lead to greater risk of drought and flood, while higher temperatures will increase water demand. On-farm management practices that target green water, can significantly alter these predictions of water shortages (Rost et al. 2009, Power 2010). Conservation agriculture improves water productivity by enhancing infiltration, reducing soil evaporation and increasing soil water holding capacity for subsequent stomatal transpiration (Dillaha et al. 2010).

- **Energy use and efficiency:** Reduction of energy requirement is possible by reducing number of tillage operations by adoption of conservation tillage. According to Kour et al. (2011), conservation tillage could offset as much as 16% of world-wide fossil fuel emissions and can slow or prevent the loss of organic C in soil. On an average, by adopting ZT for land preparation and crop establishment in rice-wheat system of the IGP, farmers could save 36 l diesel/ha (Erenstein and Laxmi 2008). Saad et al. (2016) studied the energy auditing in CA-based maize-wheat-mungbean system and found that ZT bed planting with wheat and maize residue retention could be a substitute of the conventional agricultural system for adoption in maize-wheat-greengram cropping system in the irrigated north western Indo-Gangetic Plains. Kumar et al. (2013) showed that ZT improved the operational field capacity by 81%, specific energy by 17% and the energy usage efficiency by 13% as compared to CT.

- **Preventing erosion:** Conservation agriculture and other forms of conservation tillage highly reduce soil erosion which has been documented in most of the studies regarding conservation agriculture. A study conducted in Mississippi (Dabney et al., 2004), it has found that, on a silt loam soil comparing no-till and chisel/disc till under corn cultivation, no-till decreased soil erosion significantly while the land was in no-till for 5 to 10 years and for the first year after no-till ended, but a year no-till was abandoned and tillage was reinstituted, the protective effects of the previous no-till were no longer significant.

- **Bio-diversity:** The intensification of agriculture and the traditional techniques based on extensive tillage have negative impact on soils, causing degradation in the physical and chemical parameters and causing loss of biodiversity. Conservation agriculture and the related management practices have demonstrated in the last decades to be an efficient tool to combine food productivity with environmental protection around the world. So, the reduction or no-tillage, the permanent soil cover, and the crop rotation or diversification practices have discernible positive effects on soils, including the improvement of physical and chemical properties, the decrease of water run-off and wind erosion, and an increase of water retention. The use of cover crop and the organic residues on the surface enhance the stability of soils and regulate temperatures. Therefore, the biodiversity increases (Conti 2015).

2.6. The Relationship Between Resilience and the Provision of Ecosystem Services

Productive ecosystems are necessary for the supply of ecosystem services. Ecosystems that lack resilience are vulnerable to disturbances that can lead to reductions in the supply of ecosystem services. Maintaining a stable supply of ecosystem services, therefore, depends on having resilient ecosystems. But a resilient ecosystem will allow speedy recovery of ecosystem services provision. Some ecosystem services are valuable precisely because they increase the resilience of social-ecological systems. Ecosystem services such as flood mitigation and coastal protection from storm surge reduce the size of disturbances and the destruction that floods and storm surge cause to human communities. Several studies have analyzed the value of coastal marshes, mangroves, and other habitats for protection of coastal communities from storm surge (e.g., Barbier et al., 2008; Costanza et al., 2008; Das and Vincent, 2009). It is often difficult to be precise about how much protection ecosystems are likely to provide given the variability of storms, including wind speed and direction, duration, and arrival of the storm relative to high tides (Koch et al., 2009), but there can be little doubt that their contributions can be significant. Some ecosystem services can be replaced by human-engineered services. For example, storm protection can be provided by green infrastructure (coastal marshes) or grey infrastructure (concrete barriers), and water purification for clean drinking water can be provided by forests and wetlands that filter nutrients and pollutants or by water treatment facilities. Loss of ecosystem resilience can compromise ecosystem services and may require replacement of the ecosystem service with a human-engineered service. In some instances, provision of a human engineered service may undermine the provision of other services—for example, concrete barriers could protect one section of coastline but intensify erosion in nearby locales, resulting in increased vulnerabilities elsewhere along the coast. Coastal communities are often highly dependent on coastal resources for their livelihood. The services provided by the ecosystem to the community may be quite diverse, and may include commercial and recreational fishing, beach tourism, wildlife viewing, marine transportation, and other extractive uses. Any event that impacts the ecosystem, such as an oil spill, can have an impact on multiple ecosystem services, with consequent impacts on the human community (Adger, 2000). Highly dependent communities can be severely impacted by disruptions to ecosystems that cause a loss of services.

2.7. Soil conservation and Conservation Agriculture

There is a lack of knowledge about the effectiveness and efficiency of soil conservation policies in agriculture and there is a little understanding of how policy measures should be designed to encourage farmers to adopt soil conservation practices. About one-sixth of the world's land area, that is, about one-third of the land used for agriculture, has been affected by soil degradation in the historic past. While most of this damage was caused by water and wind erosion, other forms of soil degradation are induced by biological, chemical, and physical processes. Since the 1950s, pressure on agricultural land has increased considerably owing to population growth and agricultural modernization. Small-scale farming is the largest occupation in the world, involving over 2.5 billion people, over 70% of whom live below the poverty line. Soil erosion, along with other environmental threats, particularly affects these farmers by diminishing yields that are primarily used for subsistence.

Global agricultural production basically consists of food for people, feed for livestock, fibre for industry, and fuel for energy. In fact, 95% of the agricultural output is produced on cultivation and grazing lands (Hurni et al., 1996), the rest being the products of marine ecosystems. About 11% of the surface of the world's terrestrial ecosystem is used for cultivation, another 25% is used for grazing or grass-cutting, a further 28% is covered by natural and planted forests of different qualities, and about 36% is desert land (FAOSTAT, 2005).

The world's human population has increased by a factor of 2.49 since 1950 (Worldwatch Institute, 2005), causing demand for the above-mentioned products to grow even faster due to increased dietary as well as other per capita demands. Agricultural production, on the positive side, increased even faster, for example, by a factor of 2.75 between 1950 and 1985 for grain cereals (FAOSTAT, 2005), and has remained at about this level. This early increase was due to a number of factors, including advances in agricultural research and technology, plant breeding, increased inputs in minerals and fertilizer, a slight expansion of cultivated land by a factor of 1.13 (FAOSTAT, 2005), and intensification on currently cultivated land.

At present, it appears that both the area of agricultural land and the productivity potentials have reached their upper limits, with little scope to (a) further expand agricultural land sustainably and (b) develop plants that are capable of producing even more per hectare without negatively affecting people and ecology. This, however, is contested by some scientists who claim that there are still vast areas of underutilized arable land, and others who believe that biotechnology will find additional possibilities to enhance plant and animal productivity.

In the last 50 years, pressure on soil resources has been considerably accelerated in many places, along with agricultural intensification and particularly expansion. In response, there have been efforts to conserve soil to find means of agricultural production that minimize negative impacts. Much of this reproductive activity, however, is being challenged by persistent poverty, while current actions to reduce poverty in many countries often put even greater strain on agricultural land resources.

As a response, it is recognized that Conservation Agriculture(CA) may strengthen a large set ofsoil ecosystem services (Palm et al., 2014). CA is based on three technical principles which are (i) minimum soil disturbance, (ii) retention of crop and cover crop residues and (iii) use of diversified cropping patterns (FAO, 2014;Séguy et al., 2006). CA was promoted for providing multiple beneficial effects on soil physical (Indoria et al., 2017;Patra et al., 2019), biological (Lienhard et al., 2013;Mathew et al., 2012;Souza et al., 2018) and chemical properties (Ivy et al., 2017;Parihar et al., 2018; Ranaivoson et al., 2017). Amongstsoil chemical properties, soil organic carbon (SOC) was widely studied under CA, due to its potential to adapt farming systems to climate change and variability, and mitigate the effects of climate change through carbon sequestration(Lal, 2013). Globally, these studies allowed to provide process-based analysis of the soil physical, biological and/or chemical compartments, as affected by CA practices. However, such analyses do not seem sufficient to propose consistent soil health assessments.

Soil health was defined by Kibble white et al. (2008)as the capacity of a soil to produce a good quantity and quality food and fibre together with the delivery of other ecosystem services. However, in most studies in the literature, all variables used to assess soil health are treated without proper consideration of their role along the soil health causal chain. Indeed, there is often no consideration of the cause/ consequence relation between soil properties, soil functions and soil ecosystem services (Ivy et al., 2017;Lienhard et al., 2013; Thier felder and Wall, 2012), while the realization of soil properties leads to the expression of soil functions which allows a soil to provide ecosystem services (Kibblewhite et al., 2008;Su et al., 2018). The lack of consensus on this causal chain, the different interaction between soil properties and the addition of separated soil health indicators lead to different considerations of the components of soil systems which may lead to misleading conclusions on soil functioning assessment. To fill this gap, integrative approaches based on the results of soil physical-biological-chemical interactions should be implemented (Kibblewhite et al., 2008;Vogel et al., 2018;Rinot et al., 2019). Finally, assessments should be easily transferable to smallholder farmers, development practitioners and land managers who are the first beneficiaries of soil health (Idowu et al., 2008), but most studies rely on laboratory tools, protocols, and complicated data-analysis interpretation which may limit the further dissemination of new practices (Bünemann et al., 2018).

In semi-arid and arid regions, variable pattern of rainfall and scarcity are significant constraints to agricultural productivity and here, conservation agriculture would be expected to increase infiltration, soil moisture storage and utilization, and also water use efficiency. Unfortunately, in many low rainfall regions with significant dry seasons, it is difficult to maintain an effective soil organic matter cover because of competing uses as livestock feed or sometimes as fuel. In these cases, conservation agriculture may not work because of removal of the organic cover, even with no till, results in bare soil and the formation of a soil crust, which will decrease the potency for water infiltration. This was stated by Rockström et al. (2009) in case of conservation farming strategies for arid and semi-arid agriculture in East and Southern Africa. It was seen that yields were higher with conservation agriculture than conventional tillage during drier rainy seasons but there was not much difference in yields during wetter rainy seasons suggesting that conservation farming in savannah agro-ecosystems may foremost function as a water harvesting system, which will enhance the ability of crops to check dry spells.

Conservation agriculture avoid mechanical soil disturbance to the extent possible. Avoid soil compaction beyond the elasticity of the soil. Maintain or improve soil organic matter during rotations until reaching an equilibrium level. Maintain organic cover through crop residues and cover crops to minimize erosion loss by wind and/or water.

2.8. Water Conservation and Conservation Agriculture

Water and soil nutrients remain the most limiting resources for agriculture. However, rainfed and irrigated agriculture is often hindered by the depletion of soil nutrients through surface runoff and leaching during rainy phases, as well as by the scarcity of water during dry phases. Surface runoff adversely affects the availability of water, soil nutrients, and soil organic matter for plant growth and development. Percolated water (leaching) can also affect water and nutrient availability to some extent. Agricultural activities exacerbate the removal of nutrients by either of the processes. Nutrients that are removed by surface runoff are permanently lost before reaching the root zone of the plant, while nutrients that are leached below the root zone are at least temporarily lost from the root system. Concurrently, the nutrient and water components of surface runoff and percolation may also deteriorate the water quality of wells, reservoirs, and lakes. Thus, the nutrients often removed by water movement and dynamics contribute not only to water quality deterioration, but this also imply an economic loss of soil fertility to the farmer. The current approach of agricultural systems, which promotes the use of more chemical fertilizers, particularly for vegetable production, shows a wider expansion of the above risks and is becoming a serious threat to our environment.

Hence, there is a need for a paradigm shift to improve smallholder agriculture systems that promote sustainable intensification, which encourages an increase in crop productivity with minimum inputs and protects the environment at the same time. Smallholder vegetable production may also be optimized by applying conservation agriculture (CA) practices that would improve productivity with minimum inorganic inputs and minimize adverse effects on the environment. The CA system (minimum soil disturbance, complete soil cover, and proper crop rotation) has been used to improve irrigation water use efficiency and crop productivity while controlling soil nutrient losses caused by various factors. No-tillage, despite the challenges of implementation and adoption, reduces runoff, increases percolation, and enhances water holding capacity of soils, when combined with grass mulch cover and proper crop rotation. The biological decomposition of grass mulch has improved the soil quality (adding soil nutrients) and soil structure, while a no-till practice combined with a complete soil cover reduces the soil compaction in the long term, particularly in drier regions or dry phases.

The yield of cereal crops has been increased in CA systems (due to higher water infiltration) under rainfed phases of production. Water use efficiency and the yield of vegetables significantly increase under CA in dry irrigation phases of production compared with the conventional practices. Application of CA in vegetable production systems with local varieties based on market demand has greater opportunities for adoption. in supplementary irrigation vegetable production systems, CA can provide opportunities to optimize water use by decreasing irrigation water requirements and optimize nutrient use by decreasing nutrient losses through runoff and leaching. These best management practices will not only improve yield and efficiency of inputs (water and nutrient) but also decrease pollution and protect our environment.

2.9. Biodiversity Conservation and Conservation Agriculture

Unregulated urbanization, industrialization and increased human activities have led to rapid deterioration of the environment. This has severely affected the life supporting system.

Last decade or so has seen a serious concern from people of all walks of life towards various environmental issues which include global warming, biodiversity loss, depletion of ozone layer, acid rain, dumping of hazardous wastes etc. World is a beautiful place to live in because of the variety of living organisms including plants, animals and micro-organisms with whom we share this planet. The remarkable diversity of living organisms form an inseparable and significant parts of our planet. However, the ever increasing human population and the means and methods of survival that have emerged are posing serious threats to bio-diversity. Biodiversity is a natural wealth essential for human survival as much as for the survival of other species.

In addition, biodiversity in land ecosystems generally decreases with increasing altitude. The other factors that influence biodiversity are amount of rainfall and nutrient level in soil. In marine ecosystems, species richness tends to be much higher in continental shelves. India is a country of vast diversity and it is among the 12 "mega-diversity" countries in the world. Both flora and fauna, all over the world are under an assault from a variety of indiscriminate human activities. These activities are often related to rapid growth of human population, deforestation, urbanization and industrialization.

The threat to survival or loss may be caused in three ways: Direct ways (these include Deforestation, hunting, poaching, commercial exploitation), indirect ways (these include loss or modification of the natural habitats, introduction of exotic species, pollution, etc.) and Natural causes (which include Climate change.) Among these causes, habitat destruction and over-exploitation are the main causes for this catastrophe.

Habitat (natural home) destruction: It may result from clearing and burning forests, draining and filling of wetlands, converting natural areas for agricultural or industrial uses, human settlements, mines, building of roads and other developmental projects. This way the natural habitats of organisms are changed or destroyed. These change either kill or force out many species from the area causing disruption of interactions among the species. Over exploitation reduces the size of the population of a species and may push it towards extinction.

The various benefits of biological diversity are as:

- **Protection of water resources:** Natural vegetation cover helps in maintaining hydrological cycles, regulating and stabilizing water run-off and acting as a buffer against extreme events such as floods and droughts. Vegetation removal results in siltation of dams and waterways. Wetlands and forests act as water purifying systems, while mangroves trap silt thereby reducing impacts on marine ecosystems.
- **Soil protection:** Biological diversity helps conservation of soil and retention of moisture and nutrients. Clearing large areas of vegetation cover has been often seen to accelerate soil erosion, reduce its productivity and often result in flash floods.
- **Pollution reduction:** Ecosystem and ecological processes play an important role in maintenance of gaseous composition of the atmosphere, breakdown of wastes and removal of pollutants. Some ecosystems, especially wetlands have the ability to breaking down and absorb pollutants. Natural and artificial wetlands are being used to filter effluents to remove nutrients, heavy metals, suspended solids; reduce the BOD (Biological Oxygen Demand) and destroy harmful micro-organisms. Excessive quantities of pollutants, however, can be detrimental to the integrity of ecosystems and their biota.
- **Climate stability:** Vegetation influences climate at macro as well as micro levels. Growing evidence suggests that undisturbed forests help to maintain the rainfall in the vicinity by recycling water vapour at a steady rate back into the atmosphere. Vegetation also exerts moderating influence on micro climate. Cooling effect of vegetation is a common experience which makes living comfortable. Some organisms are dependent on such microclimates for their existence.
- **Maintenance of ecological processes:** Different species of birds and predators help to control insect pests, thus reduce the need and cost of artificial control

measures. Birds and nectar–loving insects which roost and breed in natural habitats are important pollinating agents of crop and wild plants. Some habitats protect crucial life stages of wildlife populations such as spawning areas in mangroves and wetlands. Without ecological services provided by biodiversity it would not be possible to get food, pure air to breathe and would be submerged in the waste produced.

- **Food, fibre, medicines, and fuel wood and ornamental plants:** Five thousand plant species are known to have been used as food by humans. Presently about 20 species feed the majority of the world's population and just 3 or 4 only are the major staple crops to majority of population in the world. A large number of plants and animal's materials are used for the treatment of various ailments. The usage of medicinal plants in India has an ancient history, dating back to the pre-Vedic culture, at least 4000 years B. C. The therapeutic values of herbal medicines led to evolution of Ayurveda which means "science of life". It is estimated that at least 70 % of the country's population rely on herbal medicines and over 7000 species of plants are used for medicinal purposes. Wood is a basic commodity used worldwide for making furniture and for building purposes. Fire wood is the primary source of fuel widely used in third world countries. Wood and bamboo are used for making paper. Plants are the traditional source of fibre such as coir, hemp, flax, cotton, jute.
- **Future resources:** There is a clear relationship between the conservation of biological diversity and the discovery of new biological resources. The relatively few developed plant species currently cultivated have had a large amount of research and selective breeding applied to them. Many presently under-utilized food crops have the potential to become important crops in the future. Knowledge of the uses of wild plants by the local people is often a source for ideas on developing new plant products.
- **Cultural values:** Plants and animals are important part of the cultural life of humans. Human cultures have co-evolved with their environment and biological diversity can be impart a distinct cultural identity to different communities. The natural environment serves the inspirational, aesthetic, spiritual and educational needs of the people, of all cultures. In a majority of Indian villages and towns, plants like Tulsi (Ocimum sanctum), Peepal (Ficus religiosa), and Khejri (Prosopis cineraria) are planted and considered sacred and worshipped.

Conservation is the planned management of natural resources, to retain the balance in nature and retain the diversity. It also includes wise use of natural resources in such a way that the needs of present generation are met and at the same time leaving enough for the future generations. Conservation of biodiversity is important to: prevent the loss of genetic diversity of a species, save a species from becoming extinct, and protect ecosystems damage and degradation.

Conservation efforts or strategies can be grouped into the following two categories: In-situ (on-site) conservation includes the protection of plants and animals within their natural habitats or in protected areas. Protected areas are land or sea dedicated to protect and maintain biodiversity. It includes National parks & sanctuaries, Biosphere reserves, sacrization of forests and lakes. Ex-situ (off-site) conservation of plants and animals outside their natural habitats. To complement in-situ conservation efforts. Ex-situ conservation is being undertaken through setting up botanic gardens, zoos, medicinal plant parks, etc. by various agencies. The Indian Botanical Garden in Howrah (West Bengal) is over 200 years old. These include botanical gardens, zoo, and gene banks, seek bank, tissue culture and cryopreservation ("freeze preservation"). Conservation of biodiversity is like an insurance policy for future as it will ensure continuity of food chains, sustainable utilization of life support systems on earth.

Conservation agriculture (CA) can be defined by a statement given by the Food and Agriculture Organization of the United Nations as "A farming system that promotes minimum soil disturbance (i.e. No-till farming), maintenance of a permanent soil cover, and diversification of plant species. It enhances biodiversity and natural biological processes above and below the ground surface, which contribute to increased water and nutrient use efficiency and to improved and sustained crop production." As the third principle is the practicing diverse crop rotationsor crop interactions. According to an article published in *the Physiological Transactions of the Royal Society* called "The role of conservation agriculture and sustainable agriculture," crop rotation can be used best as a disease control against other preferred crops (Hobbs et al. 2007). This process will not allow pests such as insects and weeds to be set into a rotation with specific crops. Rotational crops will act as a natural insecticide and herbicide against specific crops. Not allowing insects or weeds to establish a pattern will help to eliminate problems with yield reduction and infestations within fields (FAO 2007). Crop rotation can also help build up soil infrastructure. Establishing crops in a rotation allows for an extensive build-up of rooting zones which will allow for betterwater infiltration (Hobbs et al. 2007).

2.10. Barriers to Conservation Agriculture

Conservation agriculture (CA) poses a challenge both for the scientific community and farmers to overcome the traditional mindset. Worldwide CA has spread mostly in rainfed farming system, but in India largely restricted to irrigated agriculture. The main reasons are –

i. Heterogeneous soils, high slopes and physical constraints like crusting and hard pans that reduce seed germination, decrease emergence and causes poor crop stand.

ii. Non availability of crop residues as it is used as fodder.

iii. Need for efficient in situ rain water harvesting through permanent bed and furrow system.

iv. Weed management between two crop.

A mental change of farmers, technicians, extensionists and researchers away from soil degrading tillage operations towards sustainable production systems like no tillage is necessary to obtain changes in attitudes of farmers (Derpsch, 2001). Hobbs and Govaerts (2010) however, noted that probably the most important factor in the adoption of CA is overcoming the bias or mindset about tillage. It is argued that convincing the farmers that successful cultivation is possible even with reduced tillage or without tillage is a major hurdle in promoting CA on a large scale. In many cases, it may be difficult to convince the farmers of potential benefits of CA beyond its potential to reduce production costs, mainly by tillage reductions. CA is now, considered a route to sustainable agriculture. Spread of conservation agriculture, therefore, will call for scientific research linked with development efforts. The following are a few important constraints which impede broad scale adoption of CA.

- Lack of appropriate seeders especially for small and medium scale farmers: Although significant efforts have been made in developing and promoting machinery for seeding wheat in no till systems, successful adoption will call for accelerated effort in developing, standardizing and promoting quality machinery aimed at a range of crop and cropping sequences. These would include the development of permanent bed and furrow planting systems and harvest operations to manage crop residues.
- The wide spread use of crop residues for livestock feed and fuel: Specially under rainfed situations, farmers face a scarcity of crop residues due to less biomass production of different crops. There is competition between CA practice and livestock feeding for crop residue. This is a major constraint for promotion of CA under rainfed situations.
- Burning of crop residues: For timely sowing of the next crop and without machinery for sowing under CA systems, farmers prefer to sow the crop in time by burning the residue. This has become a common feature in the rice-wheat system in north India. This creates environmental problems for the region.
- Lack of knowledge about the potential of CA to agriculture leaders, extension agents and farmers: This implies that the whole range of practices in conservation agriculture, including planting and harvesting, water and nutrient management, diseases and pest control etc. need to be evolved, evaluated and matched in the context of new systems.
- Skilled and scientific manpower: Managing conservation agriculture systems, will call for enhanced capacity of scientists to address problems from a systems perspective and to be able to work in close partnerships with farmers and other

stakeholders. Strengthened knowledge and information sharing mechanisms are needed.

2.11. Challenges in Conservation Agriculture

Conservation agriculture as an upcoming paradigm for raising crops will require an innovative system perspective to deal with diverse, flexible and context specific needs of technologies and their management. Conservation agriculture R&D (Research and Development), thus will call for several innovative features to address the challenge. Some of these are:

(a) Understanding the system: Conservation agriculture systems are much more complex than conventional systems. Site specific knowledge has been the main limitation to the spread of CA system (Derpsch, 2001). Managing these systems efficiently will be highly demanding in terms of understanding of basic processes and component interactions, which determine the whole system performance. For example, surface maintained crop residues act as mulch and therefore reduce soil water losses through evaporation and maintain a moderate soil temperature regime (Gupta and Jat, 2010). However, at the same time crop residues offer an easily decomposable source of organic matter and could harbour undesirable pest populations or alter the system ecology in some other way. No-tillage systems will influence depth of penetration and distribution of the root system which, in turn, will influence water and nutrient uptake and mineral cycling. Thus the need is to recognize conservation agriculture as a system and develop management strategies.

(b) Building a system and farming system perspective: A system perspective is built working in partnership with farmers. A core group of scientists, farmers, extension workers and other stakeholders working in partnership mode will therefore be critical in developing and promoting new technologies. This is somewhat different than in conventional agricultural R&D, the system is to set research priorities and allocate resources within a framework, and little attention is given to build relationships and seek linkages with partners working in complementary fields.

(c) Technological challenges: While the basic principles which form the foundation of conservation agriculture practices, that is, no tillage and surface managed crop residues are well understood, adoption of these practices under varying farming situations is the key challenge. These challenges relate to development, standardization and adoption of farm machinery for seeding with minimum soil disturbance, developing crop harvesting and management systems.

(d) Site specificity: Adapting strategies for conservation agriculture systems will be highly site specific, yet learning across the sites will be a powerful way in understanding why certain technologies or practices are effective in a set of situations and not effective in another set. This learning process will accelerate building a knowledge base for sustainable resource management.

(e) Long-term research perspective: Conservation agriculture practices, e.g. no-tillage and surface maintained crop residues result in resource improvement only gradually, and benefits come about only with time. Indeed, in many situations, benefits in terms of yield increase may not come in the early years of evaluating the impact of conservation agriculture practices. Understanding the dynamics of changes and interactions among physical, chemical and biological processes is basic to developing improved soil-water and nutrient management strategies (Abrol and Sangar, 2006). Therefore, research in conservation agriculture must have longer term perspectives.

2.12. Stewardship, Ecology and Conservation Agriculture

The word stewardship dates from the 15th century and combines the idea of a house or hall (*stig*), such as on an estate or large farm, with the concept of a guard (*weard*) (etymonline.com). Stewardship acknowledges that people are members of nature's household and bear responsibility for its care (Chapin et al., 2015). In this sense, stewardship reflects Aldo Leopold's (1949) land ethic that recognizes society's responsibility for and dependence upon a healthy relationship between people and nature. This ethic predates the word stewardship and is deeply rooted in philosophical and religious traditions of many cultures, including both western society (Glacken, 1983) and indigenous peoples (Berkes, 2008; Kimmerer, 2013; Nuttall et al., 2005). In an ecological context, ecosystem stewardship has been described as a framework to guide ecosystem management for desirable ecological and societal outcomes (Szaro et al., 1999). In this broad sense stewardship has much in common with adaptive ecosystem management (Chapin et al., 2009; Christensen et al., 1996; Szaro et al., 1999) and biodiversity conservation (Poiani et al., 2000; Rands et al., 2010).

Ecosystem stewardship is an action-oriented framework intended to foster the social–ecological sustainability of a rapidly changing planet. Recent developments identify three strategies that make optimal use of current understanding in an environment of inevitable uncertainty and abrupt change: reducing the magnitude of, and exposure and sensitivity to, known stresses; focusing on proactive policies that shape change; and avoiding or escaping unsustainable social–ecological traps. As we discuss here, all social–ecological systems are vulnerable to recent and projected changes but have sources of adaptive capacity and resilience that can sustain ecosystem services and human well-being through active ecosystem stewardship.

Roberts defined Stewardship or land ethic as a set of values which engenders an appreciation of and respect for the land as the basis of our prosperity and quality of life, and explained that adopting land ethic was a new concept for most Australians (1992:17). For Vanclay (1992:97), Stewardship refers to the notion that farmers are stewards of the land and that farming is a way of life that places implicit responsibility on farmers to look after the land for future generation. The

Stewardship concept recognizes that farmers may have to make uneconomical decisions in order to protect the land. The analysis of survey by Vanclay (1986) suggested that Landholders are more concerned by economic impacts of land degradation. In this survey property size was part of the explanation for the absence of the predicted linkage between land care and Stewardship. Land care respondents were significantly more likely to operate larger farms. However smaller property owners had significantly higher scores on the stewardship scale. Survey result confirmed that most of the farmer have a strong stewardship ethic and indicated that majority of landholders were concerned about the environmental impact of land degradation. A stronger mix of policy instrument is needed to support land care. e.g. wetland management is an excellent example of the mix of policy option needed to support land care.

Soil stewardship: It acts as a flagship case for the intrinsic value of natural resources and provides a valuable example for education of links between land use and health (Aidan M. Keith et al. 2016). A synergy between different separate approaches would benefit from a consolidation of relevant knowledge from the large body of existing literature, following which research gaps, or areas lacking in studies, could also be identified systematically. Studies examining the effects of land management and its change on arrange of ecosystem services are becoming familiar, but those making links to animal, human and ecosystem health are less abundant (e.g. Rhodes et al.,2013).

Aidan et al. highlighted four broad areas for consideration:

1. Relationship between soil stewardship practices and health metrics at farm and landscape scales.
2. Mediation of exposure to chemical contaminants by soil stewardship practices; contamination through the food chain (e.g. uptake by crops of veterinary pharmaceuticals via manures) and transport to air/water.
3. Mediation of the dynamics of pathogenic organisms important to crop, animal and human health by soil stewardship practices, including landscape configuration.
4. Correlations and trade-offs between ecosystem services and health metrics at landscape and regional scales.

Large scale environmental and soil surveys (e.g. Land Use/Land Cover Area Frame Survey [LUCAS], GB Countryside Survey) and research site networks (e.g. Long Term Ecological Research network), which generally have an array of collocated measures, could be exploited to help investigate such relationships. The call for soil security risk assessments by McBratney et al. (2014) could also be adapted to incorporate risks to animal, human and ecosystem health, thus encapsulating potential knock on effects to those dependent on these soils.

2.12.1. Agricultural water stewardship builds agricultural and ecological resilience:

Agricultural water stewardship can be defined as the use of water in a manner that optimizes agricultural water use while addressing the co benefits of water for food production, the environment, and human health. Agricultural water stewardship is premised on the notion that water management decisions cannot be made independent of ecological, social, and economic contexts. It takes into account local biogeographical considerations such as soil type, soil ecology, topography, agronomic inputs, and terrestrial and aquatic ecology. Agricultural water stewardship emphasizes whole farm approaches that acknowledge the role of agriculture in the local watershed and community. Effective stewardship results in the provision of multiple types of ecosystem services—the human benefits garnered from well-functioning ecosystems, including agro ecosystems. In summary, the local and regional context in which a farm operates will determine the best water management approaches to maximize water use efficiency, minimize adverse environmental impacts, and balance trade-offs for society as a whole. Agricultural water stewardship takes into account a broad range of variables in identifying balanced solutions. Importantly, it embraces the full spectrum of solutions, from cutting edge technology to the cultural and agro ecological practices that enhance the ecological integration of farming and lead to improved water quality and supply management.

2.12.2. Beneficial management practices:

Many practices that could enhance performance on a wide range of water stewardship indicators are already known, but their application should be expanded.

Examples of beneficial practices include:

On farm stewardship practices

- Use of irrigation systems that are designed for optimum – distribution uniformity, with appropriate management to also achieve a high efficiency for each irrigation and location
- Improvement of relative timing of water applications and chemical and nutrient inputs to minimize aquifer pollution
- Tail water ponds and return systems
- Vegetated filter strips
- Efforts to reinstate natural flows and hydrograph cycles
- Restoration of riparian corridors and other waterways for multiple ecosystem services (e.g., infiltration, soil and water quality, wildlife habitat, etc.)
- Soil management regimes to enhance infiltration and retention of water in soils, including cover crops, amendments, minimum tillage

- ❑ Soil moisture and crop water status monitoring
- ❑ Wind turbines for frost control
- ❑ Dry farming

Supply management

- ❑ Off stream storage ponds for irrigation and frost protection
- ❑ Managed groundwater recharge using surplus irrigation water when chemical or nutrient (e.g., nitrate) runoff is low
- ❑ Wetlands for improving water quality and wildlife habitat
- ❑ Recycling of municipal water for agricultural use, with proper attention to water quality
- ❑ Measures to better match quality of water with intended use, including drinking water for rural communities
- ❑ Modification of conveyance capabilities to move flows to recharge areas

(This list is intended to illustrate the range of possible practices and is not an exhaustive or prioritized list)

2.12.3. Fresh water facts:

1. Nearly 80 per cent of the world's population is facing threats to water security
2. More than 64 per cent of the world's wetlands have been lost in the 20th century
3. The Freshwater Living Planet Index, which measures trends in vertebrate species populations, shows a decline of 76 per cent between 1970 and 2010
4. Hydropower produces more than one fifth of the world's electricity

2.12.4. Maintenance of biological diversity:

The known effect of biodiversity on ecosystem functioning relate more strongly to traits that govern the effect of species on ecosystem processes and trait that govern the response of species to environmental variation. These known effects of biodiversity change (both loss of key species or invasion of species with large impacts)can be fostered by maintaining species that span the spectrum of effect diversity and response diversity present in the system (Elmqvist et al.2003, Suding et al. 2008).

The functional redundancy associated with species diversity provides insurance against more drastic changes in environment, such as those that may occur in the event of human mismanagement of ecosystems or change in climate. Radical changes in environment are unlikely to eliminate all species of a given functional type in a diverse ecosystem, allowing the surviving species to increase in abundance and maintain functions that might otherwise be seriously compromised.

Table 2.3: Examples of biodiversity effects on ecosystem services. (Chapin et al. modified it from Diaz et al. 2006)

Ecosystem Service	Diversity component and mechanism
1. Production by societally important plants	*Functional composition*: (a) Fast growing species produce more biomass; (b) species differ in timing and spatial pattern of resource use (complementarily allows more resource to be used) *Species number:* large species pool is more likely to contain productive species.
2. Stability of crop production	*Genetic diversity:* Buffers production against losses to pests and environmental variability. *Species number:* Cultivation of multiple species in the same plot maintains high production over a broader range of conditions. *Functional composition:* species differ in their response to environment and disturbance, stabilizing production.
3. Maintenance of soil resources	*Functional composition:* (a) fast growing species enhance soil fertility; (b) Dense root systems prevent soil erosion.
4. Regulation of water quantity and quality	*Landscape diversity:* Intact riparian corridors reduce erosion. *Functional composition*: Fast growing plants have high transpiration rates, reducing stream flow.
5. Pollination for food production and species survival	*Functional composition*: loss of specialized pollinators reduces fruit set and diversity of plant that reproduce successfully. *Species number*: Loss of pollinator species reduces the diversity of plants that successfully reproduce (genetic impoverishment) *Landscape diversity*: Large, well connected landscape units enable pollinators to facilitate gene flow among habitat patches.
6. Resistance to invasive species with negative ecological/cultural effects	*Functional composition*: Some competitive species resist the invasion of exotic species. *Landscape structure*: Roads can serve as corridors for spread of invasive species; natural habitat patches can resist spread. *Species number*: Species rich communities are likely to have less unused resources and more competitive species to resist invaders.
7.Pest and disease control	*Genetic diversity or species control*: Reduces density of suitable hosts for specialized pests and diseases. *Landscape diversity*: Provides habitat for natural enemies of pests.

Ecosystem Service	Diversity component and mechanism
8.Biophysical climate regulation	*Functional composition*: Determines water and energy exchange, thus influencing local air temperature and circulation patterns *Landscape structure*: Influences convective movement of air masses and therefore local temperature and precipitation.
9.Climate regulation by carbon sequestrations	*Landscape structure*: Fragmented landscapes have greater edge to edge ration; edges have greater carbon loss. *Functional composition*: Small, short lived plants store less carbon. *Species number*: High species number reduces pest outbreaks that cause carbon loss.
10.Protection against natural hazards (e.g., floods, hurricanes, fires)	Landscape structure: Influences disturbances spread and/or protection against natural hazards. Functional composition: (a) extensive root systems prevent erosion and uprooting; (b) deciduous species are less flammable than evergreens

2.13. Long Term Prediction in Ecology Study

Long-term ecological studies are critical for providing key insights in ecology, environmental change, natural resource management and biodiversity conservation. In this paper, we briefly discuss five key values of such studies. These are: (1) quantifying ecological responses to drivers of ecosystem change; (2) understanding complex ecosystem processes that occur over prolonged periods; (3) providing core ecological data that may be used to develop theoretical ecological models and to parameterize and validate simulation models; (4) acting as platforms for collaborative studies, thus promoting multidisciplinary research; and (5) providing data and understanding at scales relevant to management, and hence critically supporting evidence-based policy, decision making and the management of ecosystems.

There are many nuances to discussions about what constitutes 'long term' in ecological research. Strayer et al. (1986) noted that some workers consider long-term studies to be those that continue beyond the generation time of dominant organisms in an ecosystem, or sufficiently long to quantify the key processes that structure the ecosystem under investigation. Under this definition, month-long studies of bacterial assemblages with very rapid generation times could be considered long-term. Conversely, a 300-year study of stands of trees in which the dominant trees may live for 500–1000 years would not qualify as long-term. These considerations are important because they emphasize the role of variable life spans of different organisms, but they are not feasible to use for many ecosystem analyses.

Recent trends in ecological research may make it difficult to maintain long-term studies, with much emphasis now placed on approaches using meta-analysis, data mining and modelling. These can be powerful and are generally cheaper and quicker to generate trends and publications than traditional field-based, empirical ecology. This makes it difficult for field ecologists running long-term studies to compete with modellers and data summarizers when funding success is often based on publication metrics (Lindenmayer & Likens 2011). In addition, in some areas of ecological research, it appears to be increasingly difficult to publish empirical, place-based studies compared with modelling, meta-analysis and data mining. Bormann et al. (2007) and Westgate et al. (unpublished data 2011) showed that empirical examples of adaptive management are significantly harder to publish in the international peer-reviewed literature and are also significantly less cited than non-empirical opinion pieces and essays on the same topic. Yet, empirical data are the raw materials essential for intelligent parameterization of models and meta-analysis. Without ongoing support for longterm ecological studies, meta-analyses and systematic reviews will have fewer cases to review and analyse, important ecological discoveries may be overlooked and evidence-based environmental policy and management (e.g. Sutherland et al. 2004; Collaboration for Environmental Evidence 2011) will become harder to achieve (Lindenmayer & Likens 2011). Moreover, without long-term ecological studies, it will not be possible to tackle effectively key issues like the management of ecosystems for their key services and quantify and mitigate the effects of environmental change on ecosystems and on biodiversity, including climate change. To ensure the maintenance of long-term ecological studies, we believe it is important to maintain publishing outlets for empirical field-based ecology, develop new systems of recognition that reward empirical ecologists for the use of their data by others (e.g. Likens & Lindenmayer 2011; Reichman et al. 2011) and re-craft funding schemes to emphasize collaborative partnerships among field-based ecologists, theoreticians and modellers as well as between ecologists and workers in other fields as part of multidisciplinary research. Moreover, funding schemes in ecology per se need to better acknowledge the importance and manifold values of long-term studies and ensure that this translates to financial support for such work. Trends in this direction are now included in the US National Science Foundation and under the Terrestrial Ecosystem Research Network in Australia (albeit only until 2014 for the latter at this stage).

2.14. Awareness, Knowledge and Action for the Restoration of Ecosystems

The world is facing severe challenges. Billions of people around the world are suffering the consequences of the climate emergency, food and water insecurity, and the COVID-19 pandemic. Ecosystems are an indispensable ally as we meet these challenges. Protecting them and managing their resources in a sustainable manner is essential. But just increasing the protection and sustainable management of our remaining natural landscapes and oceans will not be enough: the planet's degraded ecosystems and the huge benefits that they provide must also be restored.

By declaring the UN Decade on Ecosystem Restoration, governments have recognized the need to prevent, halt and reverse the degradation of ecosystems worldwide for the benefit of both people and nature. The 2021–2030 timeline underlines the urgency of the task. Without a powerful 10-year drive for restoration, we can neither achieve the climate targets of the Paris Agreement, nor the Sustainable Development Goals.

The UN Decade on Ecosystem Restoration provides a unique opportunity to transform food, fibre and feed production systems to meet the needs of the 21st century, and to eradicate poverty, hunger and malnutrition. This we seek to achieve through effective and innovative landscapes and seascapes management that prevents and halts degradation, and restores degraded ecosystems. The restoration of forest landscapes, farming, livestock and fish-producing ecosystems should primarily contribute to restoring them to a healthy and stable state, so that they are able to provide ecosystems services and support human needs for sustainable production and livelihoods.

For example, around one third of the world's farmland is degraded, about 87 per cent of inland wetlands worldwide have disappeared since 1700, and one third of commercial fish species are overexploited. Degradation is already affecting the well-being of an estimated 3.2 billion people – that is 40 per cent of the world's population. Every single year, we lose ecosystem services worth more than 10 per cent of our global economic output.

If we can manage to reverse this trend, massive gains await us. Reviving ecosystems and other natural solutions could contribute over one third of the total climate mitigation needed by 2030. Restoration can also curb the risk of mass species extinctions and future pandemics. Agroforestry alone could increase food security for 1.3 billion people.

Restoration on a global scale requires sustained investments. But there is growing evidence that it more than pays for itself. For example, restoring coral reefs to good health by 2030 could yield an extra USD 2.5billion a year for both Mesoamerica and Indonesia; having doubled its forest cover since the 1980s, Costa Rica has seen ecotourism grow to account for 6 per cent of GDP.

While restoration science is a youthful discipline, we already have the knowledge and tools we need to halt degradation and restore ecosystems. Farmers, for instance, can draw on proven restorative practices such as sustainable farming (conservation agriculture) and agroforestry. Landscape approaches that give all stakeholders – including women and minorities – a say in decision-making are simultaneously supporting social and economic development and ecosystem health. And policy makers and financial institutions are realizing the huge need and potential for green investment.

Ecosystem restoration is the process of halting and reversing degradation, resulting in improved ecosystem services and recovered biodiversity. Ecosystem restoration encompasses a wide continuum of practices, depending on local conditions and societal choice. Depending on objectives, restored ecosystems can follow different trajectories:

- From degraded natural to more intact natural ecosystems (often by assisting natural regeneration)
- From degraded, modified ecosystems to more functional modified ecosystems (e.g. restoration of urban areas and farmlands)
- From modified ecosystems towards more natural ecosystems, providing that the rights and needs of people who depend on that ecosystem are not compromised.

2.14.1. Approaches, principles and tools for restoration:

Ecosystem restoration encompasses a wide variety of approaches that contribute to conserving and repairing damaged ecosystems (UNEP and FAO 2020). This may involve active restoration or the removal of drivers of degradation to 'passively' promote natural regeneration. Whatever the approach, restoration requires time, resources, knowledge, enabling policies and governance if it is to contribute to human well-being, economic development, climate stability and biodiversity conservation.

Many restoration projects and programmes have underperformed in the past. Adhering to standardized principles and guidelines, as well as continued monitoring and an adaptive approach, can ensure better performance and impact.

Restoration approaches:

- ***Ecological Restoration:*** Process- Assisting the recovery of a terrestrial, freshwater or marine ecosystem that has been degraded, damaged, or destroyed. Intended end point- Transition from degraded ecosystem to a reference ecosystem, which may be a natural or a cultural one.(Source: Gann et al. 2019)
- ***Forest and Landscape Restoration:*** Process- Reversing the degradation of soils, agricultural areas, forests and watersheds thereby regaining their ecological functionality.

  Intended end point ; Restoring multiple ecological, social and economic functions across a landscape and generating a range of ecosystem goods and services that benefit multiple stakeholder groups.(Source: Besseau et al. 2018)
- ***Restoration of Aquatic Production Ecosystems:*** Process- Maintaining ecosystem structure and function to support food provisioning, while minimizing impacts, rather than restoring ecosystems to an initial state before production activity started.

Intended end point- Large/oceanic marine ecosystems supporting or affected by direct and indirect impacts of fishing gears and fisheries production; recovery through changes in fishing methods and gear modification to rebuild fish stocks and reduce adverse impacts on the environment. Specificities shown for both freshwater ecosystems and coastal ones with linkages to fisheries and aquaculture. (Source: FAO 2020c)

- ❑ ***Regenerative Agriculture:*** Process- Farming that uses soil conservation as the entry point to regenerate and contribute to multiple provisioning, regulating and supporting services. Intended end point- Enhancing environmental, social and economic dimensions of sustainable food production. Soil carbon, soil health and on-farm biodiversity are restored. (Sources: Schreefel et al. 2020; Giller et al. 2021)

- ❑ ***Rewilding:*** Process- Rebuilding, following major human disturbance, a natural ecosystem by restoring natural processes and the complete or near complete food-web at all trophic levels as a self-sustaining and resilient ecosystem using biota that would have been present had the disturbance not occurred.

  Intended end point- No pre-defined end point. Functioning native ecosystems complete with fully occupied trophic levels that are natureled across a range of landscape scales. (Source: IUCN CEM, n.d.)

Principles for successful restoration

Governments do not have to choose between economic priorities and environmental concerns (Mansuy 2020), as the restoration agenda can help to deliver on goals for livelihoods, food and water security, international trade, poverty alleviation and human rights (Perring et al. 2018; CPF 2021). However, sustainable restoration requires supportive and lasting social, economic and governance structures, engaged and empowered stakeholders, and a strong scientific evidence base to guide restorative practices. Restoration as a response to ecosystem degradation needs to address its underlying causes. Achieving national restoration commitments will take not only the right policies and regulations, but also coherence across sectoral, environmental and economic policies to ensure the alignment of public incentives and consistent implementation across national and sub-subnational levels (CPF 2021). Successful restoration can involve all actors in the land or seascape, from the private sector to local and rural communities to youth. Recent research highlights 14 'meta-lessons' from seven forest landscape restoration initiatives, providing a good example of various approaches to restoration (Mansourian et al. forthcoming).

Due to social and cultural norms, women and men often hold different environmental knowledge, and thus have different priorities for restoration (Sijapati Basnett et al. 2017). Failure to incorporate a gender-responsive approach – one that ensures women and men can fully participate in and equally benefit from interventions – could worsen existing inequalities and limit the sustainability of restoration outcomes (IUCN 2017). Therefore, the Strategy of the UN Decade on Restoration calls for the development and implementation of gender-responsive methods for empowering women and girls in dialogue, planning, decision-making and implementation of ecosystem restoration (UNEP and FAO 2020).

Indigenous peoples are responsible for managing approximately 28 per cent of the world's land surface and 37 per cent of all remaining natural lands (Garnett et al. 2018). Through their traditional knowledge, customary sustainable use and collective actions in the various ecosystems that they inhabit, indigenous peoples and local communities play a vital role in both the conservation and restoration agendas (FPP 2018). Restoration plans developed in collaboration with indigenous and local communities must be equitable, ensuring that the human rights, tenure rights and resources of all stakeholders – in particular the poorest and most vulnerable – are respected. Land tenure must be clearly defined (Ding et al. 2018) and restoration efforts must ensure that the benefits generated are equitably distributed (UNEP and FAO 2020).

Such principles can help to address challenges in restoration, including effective design and implementation, accounting for complex ecosystem dynamics in the context of climate change and navigating trade-offs associated with land management priorities and decisions (Gann et al. 2019).

Guiding principles for the UN Decade on Ecosystem Restoration:

Common principles are critical for a shared vision of ecosystem restoration. Towards this end, the FAO-led Task Force on Best Practices, the Society for Ecological Restoration, and IUCN's Commission on Ecosystem Management led an expert consultation process. Based on a synthesis of published principles for restorative activities, the process resulted in nine proposed principles that underpin ecosystem restoration throughout the UN Decade and across sectors, biomes and regions. The final wording and full descriptions of these principles will be developed collaboratively during 2021 to support effective restoration efforts around the world.

- ❑ ***Principle 1:*** Promotes inclusive and participatory governance, social fairness and equity, from the start and throughout the process and outcomes.
- ❑ ***Principle 2:*** Includes a continuum of restorative activities.
- ❑ ***Principle 3:*** Aims to achieve the highest level of recovery possible for ecosystem health and human wellbeing.
- ❑ ***Principle 4:*** Addresses drivers of ecosystem degradation.
- ❑ ***Principle 5:*** Incorporates all types of knowledge and promotes their exchange throughout the process.
- ❑ ***Principle 6:*** Is tailored to the local context, while considering the larger landscape or seascape, and social-ecological and cultural settings.

- *Principle 7:*Is based on well-defined short and long-term ecological and socioeconomic objectives and goals.
- *Principle 8:* Plans and undertakes monitoring, evaluation, and adaptive management throughout the lifetime of the project or program.
- *Principle 9:* Integrates policies and measures to ensure longevity, maintain funding and, where appropriate, enhance and scale up interventions.

2.14.2. Technical and scientific developments

Recent developments have improved the ways we monitor ecosystems, identify degraded areas, work collaboratively and restore ecosystems, making restoration easier. These new methods can be employed across different levels of knowledge, resources, finances, and spatial and temporal scales. Here are three examples:

1. *Remote monitoring:* Harnessing the power of Earth observation has been instrumental in identifying degraded areas with potential for restoration, in monitoring ongoing initiatives and in developing indicators to improve restoration outcomes (Worthington et al. 2020). Earth observation data collected by satellites, aeroplanes and drones have been rapidly improving, producing datasets at high spatial and temporal resolution that enable us to monitor restoration with increased precision. Access to data is being democratized through new platforms and tools that increase both transparency and access, allowing restoration stakeholders to monitor terrestrial and marine ecosystem health in near real-time from almost anywhere in the world (Grantham et al. 2020). However, 'eyes in the sky' cannot detect all changes on the ground. Social media and other innovative technologies can offer unexpected methods for monitoring by citizens. Smartphone apps put monitoring tools into the hands of practitioners, local communities and visitors.

Monitoring the UN Decade on Ecosystem Restoration

With the aim of supporting existing targets, goals and indicators, it is key for the UN Decade to have a centralized platform for monitoring restoration and measuring progress at all scales. The FAO-led Task Force on Monitoring is working alongside UNEP, with support from the Rio Convention secretariats and experts from all eight selected ecosystems, to develop the Framework for Ecosystem Restoration (FERM). Launching in 2021 on a geospatial platform, the FERM will enable the identification of global, regional and national indicators for restoration, for all stakeholders of the UN Decade. The FERM platform will enable knowledge and technologytransfer, as well as capacity development for people, communities and countries.

1. *Better knowledge sharing:* Effective restoration is essential to the UN Decade's success. However, what works in one place may have negative consequences elsewhere. Scientific understanding of restoring ecosystems and the complex interactions between habitats, people and climate has been developing rapidly.

These advances must be able to reach those on the ground, and platforms such as the Forest and Landscape Restoration Mechanism, Restoration Resource Center and Global Tree Knowledge Platform bring together the latest resources, knowledge and databases on planting techniques, best practices and capacity building guidelines, maps and analysis packages, among others. Similar platforms exist for marine and freshwater ecosystems, such as NFI Fisheries Resources and Information System, Global Mangrove Alliance and Allen Coral Atlas. These platforms help practitioners make informed decisions and enhance the success of their restoration projects.

2. ***Improved on-the-ground practices:*** Low-tech, cost-effective practices to enhance the success of restoration are becoming widespread, easily accessible and scalable. For example, farmer-assisted natural regeneration and bio-reclamation techniques have been successful in restoring drylands.

Restoring millions of hectares to meet forest and landscape restoration targets will require many millions of plants, seedlings and seeds where natural regeneration will not meet restoration needs. Nurseries and seed banks, which allow for experimentation, could play a critical role in creating healthy and resilient ecosystems, particularly when planting in harsh climates and landscapes (Haase and Davis 2017; Chapman et al. 2019).

Incorporating agroforestry practices into the production of crops such as cocoa and coffee increases tree diversity and cover on plantations. It also helps improve and restore habitats for wildlife and provide carbon storage. When the correct balance between trees and crops is achieved, agroforestry has the potential to increase yields while offering alternative income streams for farmers through fruit or timber production from shade trees (Jose 2009). Finally, new platforms such as Collect Earth bring together Earth observation datasets and facilitate new ways of working together to process and analyse data, such as 'mapathons' to identify areas with potential for restoration. The El Salvador government used this approach to identify restoration opportunities in the Cerron Grande watershed (Reytar et al. 2021). These collaborative approaches help to include the voices of local communities and strengthen their capacity to take an active role in restoration.

3

Review and Citations

This chapter is an extensive survey of all past studies relevant to the field of investigation which gives a knowledge of what others have found out in the related field of study. In any scientific investigation or research a comprehensive presentation of review of literature is a statutory requirement. Its main functions are to develop a better understanding of the problem to be investigated, delineate a new area of study and avoid unnecessary duplication, decide on the tools and techniques to be adopted including developing some new ones, also to relate the present study with the previous ones by finding out the areas of agreement and disagreement.

The review of literature of the present study is based on the major concepts related to the field of investigation. As direct references on all the item are not in abundance, certain specific references along with some indirect references have been incorporated in this chapter for the purpose of meaningful use.

3.1 Ecological Resilience and Conservation Agriculture

Acharya, S.K. and Chatterjee, Riti (2021) They studied that in an era of unstable ecology and fragile environment, a sustainable cultivation approach is gaining gradual importance in place of conventional input-intensive agricultural production systems along with non-judicious use of agrochemicals and irrigated water. Day by day CA is getting more and more attention of farming community throughout the world. In 2015-2016 conservation agriculture was adopted on about 180 MHA of cropland, about 12.5% of the total global cropland, i.e. 69% more than 2008-2009 in the world.

Worstell, Jim (2020) He studied thatresilience of food systems is being tested by the COVID-19 disruption. Along with any severe disruption, collapse of some systems, innovation in others, and total reorganization of some will occur. Direct delivery of food, online farmer's markets, community supported agriculture operations (CSAs), backyard food production, expansion of seed producers and plant nurseries, and decrease in restaurant share of the food dollar with increased home cooking are some trends that may be lasting. These trends can be seen as complex adaptive systems following the adaptive cycles of all open systems. The crisis provides an opportunity to examine a model of food system resilience (CLIMATED) and apply it more broadly.

Dumas, Sarah E.; Lungu, Luke; Mulambya, Nathan; Daka, Whiteson; McDonald, Erin; Steubing, Emily; Lewis, Tamika; Backel, Katherine; Jange, Jarra; Lucio-Martinez, Benjamin; Lewis, Dale; Travis, Alexander J. (2016) The study reveals that due to highly variable rainfall and lack of education, agricultural inputs, and market access constrain many farmers of Zambia's Luangwa Valley employ short-term coping strategies that threaten ecosystem resilience. Community Markets for Conservation (COMACO) utilizes an agribusiness model to alleviate poverty and food insecurity through conservation farming, market development and value-added food production. COMACO promotes household, agricultural and ecological resilience along two strategic lines: improving recovery from shocks (mitigation) and reducing the risk of shock occurrence. This intervention therefore has the potential to improve not only producers' economic resilience, but also resilience tied to the food security and physical health of the entire community.

Shava, Soul; Krasny, Marianne E.; Tidball, Keith G. and Zazu, Cryton (2010) This paper explores how agricultural knowledges may be adapted and applied among relocated people in light of globalizing trends toward urbanization and resettlement. This contribution to the collection draws upon case studies of recollection and application of agricultural knowledge as revealed in narratives from immigrant gardeners in New York, USA, and relocated farmers in Sebakwe, Zimbabwe, where the communities draw upon their reserves of knowledge to respond to changes within their local environments. Such knowledge can serve as a source of community resilience through enabling people to sustain their livelihoods and community well-being, and thus adapt to environmental changes and displacement.

King, Christine A. (2008) This paper emphasizes on ecological and community resilience that provides a bridge between traditional agriculture and natural resource management. The paper reports on current research by the author to explore a range of systems s such as Organic Agriculture, Biodynamics, Community Supported Agriculture (CSA), Permaculture, Farmers Markets and Community Gardens and how they contribute to agri-ecological and community resilience. The paper explores how agri-ecological systems contribute to more sustainable and resilient communities, through community development processes such as relationship building, genuine participation, inclusiveness, resource mobilization and creating space for knowledge sharing. The paper concludes by comparing ecological systems models to agri-ecological systems, and suggests how ecological systems theories and concepts might contribute to thinking about the future of community-based agri-ecological resilience.

Walker, Brian; Holling, C.S; Carpenter, Stephen R. and Kinzig, Ann (2004) Their study is based on three related attributes of social– ecological systems (SESs) determine their future trajectories: resilience, adaptability, and transformability. The implications of this interpretation of SES dynamics for sustainability science include changing the focus from seeking optimal states and the determinants of maximum sustainable yield (the MSY paradigm), to resilience analysis, adaptive resource management, and adaptive governance (process of creating adaptability and transformability in SESs).

Gunderson, Lance H. (2000) This review is divided into three parts. The first section reviews concepts and multiple meanings of resilience as they have appeared in the literature. That section reviews examples of modeling and field experiments that enrich our understanding of ecological change. The second section includes an assessment of how resilience is related to other key ecosystem properties. The review concludes with a section on how ecological resilience is key to management of complex systems of people and nature.

Adger, W. Neil (2000) The study defines social resilience as the ability of groups or communities to cope with external stresses and disturbances as a result of social, political and environmental change. The origins of this interdisciplinary study in human ecology, ecological economics and rural sociology are reviewed, and a study of the impacts of ecological change on a resource-dependent community in contemporary coastal Vietnam in terms of the resilience of its institutions has been outlined.

Peterson, Garry; Allen, Craig R. and Holling, C. S. (1998) Their study described theexisting models of the connection between species diversity and ecological function, and proposed a conceptual model that relates species richness, ecological resilience, and scale. A suggestion that species interact with scale-dependent sets of ecological structures and processes that determine functional opportunities. A proposal of ecological resilience is generated by diverse, but overlapping, function within a scale and by apparently redundant species that operate at different scales, thereby reinforcing function across scales is given. The distribution of useful diversity among and across scales allows regeneration and renewal to occur following ecological disruption over a good vary of scales.

3.2 Ecosystem Services and Conservation Agriculture

Ghosh, Sonaka; Das, T K; Sharma, D K and Gupta, Kamlika (2019) Their study explainedthat Conservation agriculture (CA) has emerged as a promising technology for efficient rational use of available resources and sustained productivity in the long run. By saving inputs, reducing energy usage and greenhouse gases emissions, CA-based management practices are quite viable for bringing sustenance in agricultural crop production. The paper focus on the potential ecosystem service benefits accrued from CA. Conservation agriculture in the long run can be a strategy for sustainable crop intensification and a climate resilient crop management system.

Acharya, S.K. and Chatterjee, Riti (2019) They found in our operating agro-ecosystem, the provider of food-fuel-fodder, water is contaminated, soil is eroded, organic carbon is depleted and biodiversity is marginalized. So, we need to have

Conservation Agriculture with a goal to protect and preserve our precious natural resources, stabilizing and augmenting food production and livelihood for millions.

Garbach, Kelly; Milder, Jeffrey C.; DeClerck, Fabrice A.J.; Montenegro de Wit, Maywa; Driscoll, Laura and Gemmill-Herren, Barbara (2016) This review develops an analytic framework to characterize the fulfilment of these objectives by documenting the co-occurrence of positive, neutral, and negative outcomes for crop yield and nine regulating ecosystem services. The paper provides an illustrative examination of the framework, evaluating evidence for yield and ecosystem service outcomes across five AEI systems: conservation agriculture, holistic grazing management, organic agriculture, precision agriculture, and system of rice intensification (SRI). The review presents substantial evidence that the five AEI systems can contribute to multifunctional agriculture by increasing ecosystem service provision, or reducing negative externalities associated with agriculture, while maintaining or increasing yields.

Giller, Ken E.; Andersson, Jens A; Corbeels, Marc; Kirkegaard, John; Mortensen, David; Erenstein, Olaf and Vanlauwe, Bernard (2015) They have found that over the past 10 years CA has been promoted among smallholder farmers in the (sub-) tropics, often with disappointing results. Growing evidence challenges, the claims that CA increases crop yields and builds-up soil carbon although increased stability of crop yields in dry climates is evident. Their analyses suggested a pragmatic adoption on larger mechanized farms, and limited uptake of CA by smallholder farmers in developing countries. They have also proposed a rigorous, context-sensitive approach based on Systems Agronomy to analyse and explore sustainable intensification options, including the potential of CA.

Farooq, Muhammad and Siddique, Kadambot H. M. (2014) They found CA involves changing many conventional farming practices as well as the mindset of farmers to overcome the conventional use of tillage operations. Although adoption of CA is increasing globally, in some regions it is either slow or non-existent. The adoption of CA has both agricultural and environmental benefits but there is a lack of information on the effects and interactions of key CA components which affect yield and hinder its adoption. This paper also discusses about the basic concepts and brief history of CA, and its impacts on agricultural systems.

Palma, Cheryl; Blanco-Canqui, Humberto; DeClerck, Fabrice; Gaterea, Lydiah and Grace, Peter (2014) Their study reveals that Conservation agriculture (CA) changes soil properties and processes compared to conventional agriculture. These changes can, in turn, affect the delivery of ecosystem services, including climate regulation through carbon sequestration and greenhouse gas emissions, and regulation and provision of water through soil physical, chemical and biological properties. Conservation agriculture can also affect the underlying biodiversity that supports many ecosystem services. In this overview, we summarize the current status of the science, the gaps in understanding, and highlight some research priorities for ecosystem services in conservational agriculture. The review is based on global literature but also addresses the potential and limitations of conservation agriculture for low productivity, smallholder farming systems, particularly in Sub Saharan Africa and South Asia.

Sanderson, Matt A.; Archer, David; Hendrickson, John; Kronberg, Scott; Liebig, Mark; Nichols, Kris; Schmer, Marty; Tanaka, Don and Aguilar, Jonathan (2013) In this paper, they explored the concept of diversity and its role in maximizing ecosystem services from managed grasslands and integrated agricultural systems (i.e., integrated crop–livestock–forage systems) at the field and farm level. They also examined trade-offs that may be involved in realizing greater ecosystem services. Their study also reveals that integrating dynamic cropping systems with livestock production increases the complexity of management, but also creates synergies among system components that may improve resilience and sustainability while fulfilling multiple ecosystem functions. Diversified conservation agricultural systems can sustain crop and livestock production and provide additional ecosystem services such as soil C storage, efficient nutrient cycling and conservation of biodiversity.

Power, Alison G. (2010) Their study emphasizes on agricultural ecosystems which rely on ecosystem services provided by natural ecosystems, including pollination, biological pest control, maintenance of soil structure and fertility, nutrient cycling and hydrological services. Agroecosystems also produce a variety of ecosystem services, such as regulation of soil and water quality, carbon sequestration, support for biodiversity and cultural services. The study focuses on the tradeoffs that may occur between provisioning services and other ecosystem services and disservices should be evaluated in terms of spatial scale, temporal scale and reversibility. As more effective methods for valuing ecosystem services become available, the potential for 'win–win' scenarios increases. Under all scenarios, appropriate agricultural management practices are critical to realizing the benefits of ecosystem services and reducing disservices from agricultural activities.

Nelson, Erik; Mendoza, Guillermo; Regetz, James; Polasky, Stephen; Tallis, Heather; Cameron, D Richard; Chan, Kai MA; Daily, Gretchen C; Goldstein, Joshua; Kareiva, Peter M; Lonsdorf, Eric; Naidoo, Robin; Ricketts, Taylor H and Shaw, M Rebecca (2009) They have observed that there is increasing consensus about the importance of incorporating "ecosystem services" into resource management decisions, but quantifying the levels and values of these services has proven difficult. They used a spatially explicit modeling tool, Integrated Valuation of Ecosystem Services and Tradeoffs (InVEST), to predict changes in ecosystem services, biodiversity conservation, and commodity production levels. They have also applied InVEST to stakeholder defined scenarios of land-use/land-cover change in the Willamette Basin, Oregon where they found that scenarios that received high scores for a variety of ecosystem services also had high scores for biodiversity, suggesting there is little tradeoff between biodiversity conservation and ecosystem services.

3.3 Stewardship, Ecology and Conservation Agriculture

Chapin, F Stuart; Carpenter, Stephen R.; Kofinas, Gary P.; Folke, Carl; Abel, Nick; Clark, William C.;Olsson, Per; Smith, D. Mark Stafford; Walker, Brian; Young, Oran R.; Berkes, Fikret; Biggs, Reinette; Grove, J. Morgan; Naylor, Rosamond L.; Pinkerton Evelyn; Steffen, Will; Swanson, Frederick J. (2010) The study reveals that ecosystem stewardship is an action-oriented framework intended to foster the social–ecological sustainability of a rapidly changing planet. Recent developments identify three strategies that make optimal use of current understanding in an environment of inevitable uncertainty and abrupt change: reducing the magnitude of, and exposure and sensitivity to, known stresses; focusing on proactive policies that shape change; and avoiding or escaping unsustainable social–ecological traps. As discussed here, all social–ecological systems are vulnerable to recent and projected changes but have sources of adaptive capacity and resilience that can sustain ecosystem services and human well-being through active ecosystem stewardship.

Foster, David; Swanson, Frederick; Aber, John; Burke, Ingrid; Brokaw, Nicholas; Tilman, David and Knapp, Alan (2003) Their study demonstrates the ubiquity and importance of land-use legacies to environmental science and management and recognizes that the legacies of land-use activities continue to influence ecosystem structure and function for decades or centuries—or even longer—after those activities have ceased. Consequently, recognition of these historical legacies adds explanatory power to our understanding of modern conditions at scales from organisms to the globe and reduces missteps in anticipating or managing for future conditions. As a result, environmental history emerges as an integral part of ecological science and conservation planning.

Hails, Rosemary S. and Ormerod, Steve J. (2013) They observed that national and international assessments are increasingly highlighting the unsustainable use of earth's natural resources in the face of population increase, growing material affluence and global change. In contrast to resource depletion, the concept of natural capital emphasizes how the environment is an asset to be managed, to ensure that the benefits which flow from it are sustained for future generations. There are specific challenges to ecological science in this interdisciplinary endeavour: specifically, to develop frameworks for identifying and monitoring natural capital; to parameterise factors affecting ecosystem services and their resilience to change; to integrate the complexity of ecological systems into ecosystem service valuation; and to characterise the synergies and trade-offs between ecosystem services in different management and policy scenarios.

Mathevet,Raphaël; Bousquet, François; Larrère, Catherineand Larrère, Raphaël(2018) This paper explores and discusses the various meanings of the stewardship concept in the field of sustainability science. The authors highlighted the increasing differences between alternative approaches to stewardship and propose a typology to enable scientists and practitioners to more precisely identify the basis and objectives of the concept of stewardship. At first they presented the two dimensions which were used to map the diversity of stances concerning stewardship. Secondly, they analysed these positions in relation to the limits of the systemic approach, ideological manipulation, responsibility, and solidarity. In the final section they explained how the concept of ecological solidarity, a core principal in recent French law on biodiversity conservation and national park governance can contribute to the underpinning of a specific form of social-ecological stewardship.

Hobbs, Peter R. (2007) This paper focuses on conservation agriculture (CA), defined as minimal soil disturbance (no-till) and permanent soil cover (mulch) combined with rotations, as a more sustainable cultivation system for the future. The paper first introduces the reasons for tillage in agriculture and discusses how this age-old agricultural practice is responsible for natural resource and soil degradation. The paper goes on to introduce conservation tillage (CT), a practice that was borne out of the American dust bowl of the 1930s, before comparing CT with CA, a suggested improvement on CT, where no-till, mulch, and rotations significantly improve soil properties (physical, biological, and chemical), and other biotic factors, and enables more efficient use of natural resources. Recent data is presented showing how CA can improve agriculture through improvement in water infiltration and reduced erosion, improved soil surface aggregates, reduced compaction through promotion of biological tillage, increased surface soil organic matter and carbon content, moderated soil temperatures, and weed suppression. The paper concludes that agriculture in the next decade will have to sustainably produce more food from less land through more efficient use of natural resources and with minimal impact on the environment in order to meet growing population demands.

Worrell, Richard and Appleby, Michael C. (2000) Their study explains stewardship as a potentially useful concept in modernizing management philosophies. Use of the term has increased markedly in recent years, yet the term is used loosely and rarely defined in land management literature. The connections between this practical usage and the ethical basis of stewardship are currently poorly developed. The following definition has been proposed: "Stewardship is the responsible use (including conservation) of natural resources in a way that takes full and balanced account of the interests of society, future generations, and other species, as well as of private needs, and accepts significant answerability to society."

3.4 Erosion of Natural Resources and Issues of Ecological Resilience

Knickel, Karlheinz (1990) The author observed thatcurrent farm practices contribute to a wide array of environmental problems: the monoculture of many agricultural regions, the problem of soil erosion, the contamination of ground water resources by nitrate and pesticides, wildlife habitats lost or subject to fragmentation and highlighted that environmental policies need to be an integral part of regional development and agricultural policies.

Whitlow, Richard (1988) This paper describes contemporary patterns of erosion in Zimbabwe on the basis of detailed examination of nearly 8500 aerial photographs. It discusses the main factors influencing land degradation, showing that population density and farming systems dominate the physical risks of erosion as key determinants of degradation. Conservation history in commercial and peasant farming areas is reviewed as a basis for understanding the current status of erosion. Finally, aspects of a future conservation strategy are outlined in terms of political, educational, agrarian and research issues.

Wilkinson, Cathy (2011) The study makes a theoretical contribution to this emerging inter-disciplinary exploration. It engages the following questions – what, if any, new conceptual ground does social-ecological resilience offer planning theory, and more broadly what issues does social-ecological resilience raise for further scholarship by planning theorists? The study mainly explores what new conceptual ground social-ecological resilience offers planning theory.

Nichols, James D.; Koneff, Mark D.; Heglund, Patricia J.; Knutson, Melinda G.; Seamans, Mark E.; Lyons James E.; Morton, John M.; Jones, Malcom T.; Boomer, G.Scott and Williams, Byron K. (2011) They observed that climate change and its associated uncertainties are of concern to natural resource managers. Although aspects of climate change may be novel (e.g., system change and nonstationarity), natural resource managers have long dealt with uncertainties and have developed corresponding approaches to decision-making. Their study encompasses on adaptive resource management which is an application of structured decision-making for recurrent decision problems with uncertainty, focusing on management objectives, and the reduction of uncertainty over time.

Cross, Molly S.; McCarthy, Patrick D.; Garfin, Gregg; Gori, David and Enquist, Carolyn A. F. (2012) In this paper the authors presented a model for collaborative planning aimed at identifying ways to adapt management actions to address the effects of climate change in landscapes that cross public and private jurisdictional boundaries. Continued investment in adaptation of management plans and actions to address the effects of climate change in the south-western United States and extension of the approaches used in this project to maintain the biological diversity and ecosystem services in a rapidly changing world.

Nearing, Mark A.; Xie, Yun; Liu, Baoyuan and Ye, Yu (2017) They observed that the regions of land that are brought into crop production from native vegetation typically undergo a period of soil erosion instability, and long term erosion rates are greater than for natural lands as long as the land continues being used for crop production. Average rates of soil erosion under natural, non-cropped conditions have been documented to be less than 2 Mg ha^{-1} yr^{-1}. On-site rates of erosion of lands under cultivation over largecroplandareas, such as in the United States, have been documented to be on the order of 6 Mg ha^{-1} yr^{-1} or more. In north-eastern China, lands that were brought into production during the last century are thought to have average rates of erosion over this large area of as much as 15Mg$ha^{-1}yr^{-1}$or more. Broadly applied soil conservation practices, and in particularconservation tillageand no-till cropping, have been found to be effective in reducing rates of erosion, as was seen in the United States when the average rates of erosion on cropped lands decreased from on the order of 9Mg$ha^{-1}yr^{-1}$to 6 or 7Mg$ha^{-1}yr^{-1}$between 1982 and 2002, coincident with the widespread adoption of new conservation tillage and residue management practices.

3.5 Pollution & Erosion due to Modernization of Agriculture

Bharti, Chandrabhan; Mohapatra, Anita; Maurya, Rakesh; Maharana, Chandan; Maurya, Alok and Malakar, Pranab (2020) They studied that nitrate formed by biochemical activities of microorganisms or addition of chemically synthesized forms in the lithosphere and biosphere that enters hydrosphere with relative ease, as all these environmental components are dynamically interconnected. Balancing the amount of N needed for optimum plant growth while minimizing the NO3- that is transported to ground and surface waters remains a major challenge for everyone attempting to understand and improve agricultural nutrient use efficiency. Strategies for reducing NO3- loss through drainage include improved timing of N application at appropriate rates, using soil tests and plant monitoring, diversifying crop rotations, using cover crops, reducing tillage, optimizing N application techniques, and using nitrification inhibitors. Maximum permissible limits of nitrate concentration in water prescribed by ICMR is 10 as NO3-N (mg/l) and 45 as NO3- (mg/l). The production of food grains has increased by 2% in the last 5 years as against an 8.6% rise in population.

Ahmed, Saleh and Cokinos, Christopher (2017) In this paper the authors described Bangladesh as the world's most climate-vulnerable country; on the other hand, it is also a champion of various innovative adaptation initiatives. Innovations and diffusions of agricultural technology are rapidly expanding as part of nation's effort to increase its agricultural productivity. Ecological Modernization Theory can be used to explain this situation as a response to climate-related risks and in terms of the need to achieve sustainability. This article analyses the changing pattern of climate–society interactions and advances our understanding of the human dimensions of climate change impact at the local level. Even though it has a regional focus on southwest coastal Bangladesh, the insights here – concerned with the water–energy–food–climate nexus, the local evolution of EM, and the ethical dimensions of food and agriculture – are relevant in other parts of the world that face changing patterns of climate–society interactions under limited-resource conditions.

Awasthi, Amit; Singh, Nirankar; Mittal, Susheel; Gupta, Prabhat K.; Agarwal, Ravinder (2010) They studied the variations in pulmonary function tests (PFTs) due to agriculture crop residue burning (ACRB) on children between the age group of 10 to 13 years and the young between 20 to 35 years. The effects of exposure to smoke due to rice–wheat crop residue burning on pulmonary functions like Force Vital Capacity (FVC), Force Expiratory Volume in one second (FEV1), Peak Expiratory Flow (PEF) and Force Expiratory Flow in 25 to 75% of FVC (FEF25–75%) on 40 healthy subjects of rural/agricultural area of Sidhuwal village of Patiala City were investigated for a period from August 2008 to July 2009.Small size particulate matter (PM2.5 and PM10) affected the PFTs to a large extent in comparison to the large size particulate matter (SPM). The study indicates that ACRB is a serious environmental health hazard and children are more sensitive to air pollution, as ACRB poses some unrecoverable influence on their PFTs.

Moss, Brian (2007) His study reveals that agriculture disrupts all freshwater systems hugely from their pristine states. The former reductionist concept of pollution was of examining individual effects of particular substances on individual taxa or sub-communities in freshwater systems, an essentially ecotoxicological concept. The European Water Framework Directive implicitly recognizes this in requiring restoration of water bodies to 'good ecological quality', which is defined as only slightly different from pristine state. The implications for the management of agriculture are far more profound than is currently widely realized.

Vatn, Arild; Bakken, Lars; Bleken, Marina A.; Baadshaug, Ole Hans; Fykse, Haldor; Haugen, Lars E.; Lundekvam, Helge; Morken, John; Romstad, Eirik; Rørstad, Per Kristian; Skjelva˚g, Arne O.; Sogn, Trine (2006) This paper depicts a methodology for analyzing the effect of policies focused at reducing pollution from agriculture. Such a methodology must take into account that agricultural pollution is an effect of a large set of interacting processes, covers many different substances, and may vary substantially due to shifts in natural and economic conditions. This methodology is founded on the idea of partitioning. It implies structuring and simplifying existing variation in space and time into partitions that are considered homogeneous. These partitions are organized in a hierarchy, and the different processes involved are modeled at the relevant level. The authors have concluded that analyses with fairly high level of resolution are preferable. This way it is also possible to combine a systems perspective with disciplinary integrity. A modeling structure – ECECMOD (2.0) – based on the developed principles is documented. The paper also shows the ability of this structure to simulate choice of farming practices and emissions that are well in accordance with observations from four Norwegian regions with very different agricultural and natural conditions.

Saysel, Ali Kerem; Barlas, Yaman and Yenigun, Orhan (2002) This paper examines the potential long-term environmental problems of the South-eastern Anatolian Project (GAP) related to water resources, land use, land degradation, agricultural pollution and demography are analysed from a systems perspective.

The analysis focuses on the totality of environmental, social and economic issues. For this purpose, a system dynamics simulation model (GAPSIM) has been developed as an experimental platform for policy analysis. GAPSIM was validated, first 'structurally', using the tests suggested by the literature and then the model 'behaviour' was tested and calibrated with respect to available data. The reference behaviour of GAPSIM reveals that, as the irrigated lands are developed, GAP faces significant water scarcity because of the increased intensity of cotton, the crop with the highest demands for water. GAPSIM promises to be not only a useful laboratory for policy makers of GAP, but also a useful generic structure applicable to other similar regional development projects.

Chamala, Shankariah (1990) This paper examines the external megaforces, namely international agricultural centres, trade, bilateral aid, and international organizations affecting modernization of agriculture in developing countries, and provides a balanced critique on positive and negative impacts of modernization in these countries. General trends of social and environmental impacts are briefly summarized. New approaches undertaken by international centres and by some developing countries to overcome social and environmental impacts of modern agriculture are reviewed. Mechanisms to generate appropriate agricultural technology development and management of delivery services are suggested to achieve integrated socio-economic welfare of the people and sustainable environment.

3.6 Water Conservation and Conservation Agriculture

Belay, Sisay A.; Assefa, Tewodros T.; Prasad, P. V. Vara; Schmitter, Petra; Worqlul, Abeyou W.; Steenhuis, Tammo S.; Reyes, Manuel R. and Tilahun, Seifu A. (2020) Their study aimed at investigating the impacts of conservation agriculture on irrigation water use, nutrient availability in the root zone, and crop yield under supplementary irrigation. In this study, conservation agriculture (CA), which includes minimum soil disturbance, grass mulch cover, and crop rotation, was practiced and compared with conventional tillage (CT). The authors used two years' (2018 and 2019) experimental data under paired-t design in the production of a local variety green pepper (Capsicum annuum L.). This research improves our understanding of water and nutrient dynamics in green pepper grown under CA and CT. Use of CA provides opportunities to optimize water use by decreasing irrigation water requirements and optimize nutrient use by decreasing nutrient losses through the runoff and leaching.

Pradhan, Aliza; Chan, Catherine; Roul, Pravat Kumar; Halbrendt, Jacqueline; Sipes, Brent (2017) Their study discusses about the precarious food security situation which is especially dangerous in the central Indian tribal belt (also known as the poverty belt), a typical rainfed area dominated by tribal communities. Small land holdings and their low productivity, along with uncertainties in weather patterns, degraded soils increases economic and social risks for these farmers. Under such conditions, one approach to achieve improved crop production is to minimize soil and other natural resource degradation by adopting a set of crop-nutrient-water-land system management practices, such as conservation agriculture (CA). To assess the effect of introduced technology under local ecological and socio-economic conditions, the study focused on two ecosystem services: a) provisional, and b) regulatory through five treatments consisting of farmers' traditional practice (FP) which was conventional tillage with broadcast of local variety maize (Zea mays L.); and four CA treatments viz., conventional tillage with sole cropped maize using line sowing of the improved maize cultivar 'Nilesh' (CT-M); conventional tillage with maize intercropped with the improved cowpea (Vigna unguiculata L. cultivar 'Hariyalli Bush') (CT-M + C); reduced tillage with sole cropped maize (MTM); and reduced tillage with maize + cowpea (MT-M + C). The results clearly demonstrate the potential of CA to simultaneously increase yield, diversify crop production and improve soil quality which should support a move towards sustainable intensification of crop production to improve future household income and food security. Additionally, using a transdisciplinary approach fully engaged all stakeholders in co-designing the CA treatments appropriate for the farmers and local environmental conditions leading to significant impacts on economic livelihoods, environmental sustainability and food security.

Lal, Rattan (2015) observed strategies how to use CAjudiciously and prudently to harness its potential as a conservation-effective technology, climate-resilient agriculture, and a viable option for sustainable intensification of agroecosystems for advancing food and nutritional security, and for adaptation/mitigation of climate change.Plow-based agriculture has exacerbated the problems of accelerated soil erosion by water and wind, oxidation of SOM, and decline in soil structure (aggregation) and tilth. As anthropogenic greenhouse gas (GHG) emissions increase due to global economic and population growth and the atmosphere and oceans warm, there is a strong need to identify potential C sinks of storing atmospheric CO2.

Maggioni, Elena (2015) The study reveals that Southern California is subject to long droughts and short wet spells. Its water agencies have put in place voluntary, mandatory, and market-based conservation strategies since the 1980s. By analyzing water agencies' data between 2006 and 2010, this research studies whether rebates for water efficient fixtures, water rates, or water ordinances have been effective, and tests whether structural characteristics of water agencies have affected the policy outcome. It finds that mandates to curb outdoor water uses are correlated with reductions in residential per capita water usage, while water rates and subsidies for water saving devices are not. It also confirms that size is a significant policy implementation factor. Three factors could improve the conservation effort: using prices as a conservation tool, not only as a cost recovering instrument; investing in water efficient tools only when they provide significant water savings; supporting smaller agencies in order to give them opportunities to implement conservation strategies more effectively or to help them consolidate.

Kassam, A.; Derpsch, R. and Friedrich, T. (2014) This article is an overview of achievements in soil and water conservation on agricultural lands through the experience derived from the adoption and spread of Conservation Agriculture (CA) world-wide. CA is an agro-ecological approach to sustainable production intensification. It involves the application of three inter-linked principles: (i) permanent no or minimum mechanical soil disturbance, which in practice entails direct seeding through mulch into no-till soils; (ii) maintenance of soil cover with crop residues and green manure crops, particularly legumes; and (iii) diversified cropping system involving annuals and perennial in rotations, sequences and associations. In 2011, CA had spread over 125 million hectares (9% of the global cropped land) across all continents and most agro-ecologies, including small and large farms. In addition, there is a significant area of CA orchards in the Mediterranean countries. CA is now considered to be a practical agro-ecological approach to achieving sustainable agriculture intensification. It offers environmental, economic and social advantages that are not fully possible with tillage-based production systems, as well as improved productivity and resilience, and improved ecosystem services while minimizing the excessive use of agrochemicals, energy and heavy machinery.

Andersson, Jens A.; D'Souza, Shereen (2013) Their study is the review of Conservation Agriculture (CA) adoption among smallholder farmers in southern Africa (Malawi, Zambia and Zimbabwe) analyses the historical background of the upsurge in CA promotion, the various definitions of CA that have emerged since the 1990s, the barriers to its adoption, as well as uptake figures and adoption studies. A more thorough analysis of farming households and their resource allocation strategies is required to understand the farm-level adoption constraints different types of farmers face. As contextual factors appear key influences on smallholders' farming practices, studies focusing on the wider market, institutional and policy context which are needed to understand (limited) CA adoption in Southern Africa.

Agar, John W. M.; Simmonds, Rosemary E.; Knight, Richard and Somerville, Christine A. (2009) Their study focusses on resource conservation, most hemodialysis (HD) services that wastefully or ignorantly discard reverse osmosis (R/O) "reject water" (RW) to the sewer. An R/O system is producing the highly purified water necessary for dialysis, it rejects any remaining dissolved salts from water already prefiltered through charcoal and sand filters in a high-volume effluent known as RW. The authors believe the RW recycling and re-use is feasible, simple, safe, and effective. It is community-minded, environment tally responsible, and inexpensive to establish. Once operational, ongoing reductions in water use are profound and, furthermore, the goodwill this extends to a highly water-conscious community such as our own is inestimable. They also believe that our precious resources should not be wasted any longer.State and federal systems should mandate water conservation practices in dialysis services both nationally and internationally as part of a heightened recognition of natural resource conservation.

Sarangi, A.; Madramootoo, C. A. and Cox, C. (2004) The prime objective of this work has been to develop a Decision Support System (DSS) on watershed conservation measures incorporating the established information on implementation of structural and vegetative control measures based on slope, soil type, Land Capability Classes and sediment-loss data. Conceptual decision flow logic was formulated to link this information through the Visual Basic programming language. The DSS was developed with graphic user interface (GUI) capability. This provides the user with an interactive environment to enter input information of slope, soil type, LCC and sediment loss, and then to obtain structural and vegetative control measures in response to those input values. These suggested measures displayed in decision boxes vary according to the input values as per the decision-flow logic used in DSS development.

3.7. Soil Conservation and Conservation Agriculture

Du, Ling; McCarty, Gregory W.; Li, Xia, Rabenhorst, Martin C.; Wang, Qianfeng; Lee, Sangchul; Hinson, Audra L. and Zou, Zhenhua (2021)suggested that accurate prediction of soil organic carbon (SOC) at fine scales is frequently limited by data availability but fine-scale maps of soil properties are often of great importance for local agricultural land management and watershed processes assessment. Terrain attributes derived from elevation have been commonly used predictor variables in SOC mapping. The aim of this investigation is to plan SOC thickness (0–30 cm) utilizing cross breed strategies that consolidate territorial topographic models and a remaining kriging methodology utilizing local samples from an area of interest and assess the impacts of various sampling designs for residual kriging on final map quality. This examination exhibits the utility of extrapolating a territorial topographic model by utilization of a limited number of new examples for local alignment and gives understanding to specialists who are keen on mapping soil carbon distribution in sparsely sampled areas of interest.

Busari, Mutiu Abolanle; Kukal, Surinder Singh; Kaur, Amanpreet; Bhatt, Rajan and Dulazi, Ashura Ally (2015) This paper aims to review the work done on conservation tillage in several agro-ecological regions thus on perceive its impact from the views of the soil, the crop and the environment. The reports from the study have known many edges of conservation tillage over standard tillage (CT) with reference to soil physical, chemical and biological properties similarly as crop yields. Conservation tillage involving ZT and minimum tillage that has potential to interrupt the surface compact zone in soil with reduced soil disturbance offers to steer to a more robust soil surroundings and crop yield with least impact on the surroundings.

Prager, K.; Schuler, J.; Helming, K.; Zander, P.; Ratinger, T. and Hagedorn, K. (2011) conducted a study on institutional settings close agricultural soil management in 10 European countries supported the establishments of property (IOS) framework. This framework considers the interdependencies between ecological and social systems, taking into consideration environmental conditions, farming practices impacting on conservation, differing types of actors, policies, institutions and governance structures. The aim of this paper is to elucidate the analytical framework and thus the methodology that each one case studies are supported with present and discuss compared findings, outline implications for winning conservation policy and draw conclusions on the tactic approach.

Thierfelder, C. and Wall, P. C. (2010) conducted a study to show clear advantages of CA practices with residue retention and crop rotation over conventionally tilled practices. As we have a tendency to discovered no serious pests or diseases within the trial, the study highlighted that were more advantages of crop rotations than simply pests and disease reduction. All soil quality indicators measured showed positive responses to residue retention and reduced tillage. Conservation agriculture plots generally possessed higher populations of earthworms, higher total carbon and more water stable aggregates and better infiltration rates were, however, solely found in some rotations and more available soil water on CA plots solely in some years.

Fabietti, Gabriele; Biasioli, Mattia; Barberis, Renzo and Ajmone-Marsan, Franco (2009) The purpose of this work was to supply a summary of the diffuse soil contamination on a regional scale for an outsized set of contaminants. Soil inorganic (Cd, Cr, Ni, Pb, Cu, Zn) and organic contaminants (polychlorinated dibenzo-p-dioxins(PCDD), polychlorinated biphenyls (PCBs), and polycyclic aromatic hydrocarbons (PAHs) likewise as different soil general parameters were investigated on 18×18 km grid covering an entire Italian region heavily industrial and intensively cultivated. Soils were sampled at completely different depths each for natural–undisturbed soil and for agricultural–plow soil. Variations within the trends of investigated contaminants, as a consequence of land use, vertical variability, natural or geogenic origin, relationships among contaminants, and with main soil properties, were explored by means that of enrichment factors, bi- and multivariate statistics. though natural soils during this study square measure largely situated faraway from major sources of contamination, notable variations appeared compared to agricultural areas. Contaminants like PCDDs, PCBs, PAHs, and Pb conferred high vertical variability, confirming their anthropogenic origin and powerful affinity to soil organic matter, whereas others seem to be additional associated with the natural background. The comparison of knowledge with those from soils at intervals a large-industrial city mentioned in the study space confirms the robust enrichment of the urban surroundings with relation to some contaminants like metals, PCBs, and PAHs and the other contaminants like PCDD/DFs showed lower variations confirming their diffuse and ubiquitarious pollution.

Hurni, Hans; Herweg, Karl; Portner, Brigitte; and Liniger, Hanspeter (2008) Their study focusses on the one-third of land used for agriculture out of one-sixth of world's land area thathas been stricken by soil degradation within the historic past, where most of this injury was caused by water and wind erosion, other forms of soil degradation are induced by biological, chemical, and physical processes. Since the 1950s, pressure on agricultural land has increased considerably owing to population growth and agricultural modernization. Small-scale farming is the largest occupation in the world, involving over 2.5 billion people, over 70% of whom live below the poverty line. Soil erosion along with alternative environmental threats, notably affects these farmers by decreasing yields which were primarily used for subsistence. Therefore, soil and water conservation measures are the need of the hour. Local and science-based innovations are available for most agroecological conditions and land management and farming types. Principles and measures developed for small-scale as well as modern agricultural systems have begun to show positive impacts in most regions of the world, particularly in wealthier states and modern systems. Much more emphasis still needs to be given to small-scale farming, which needs external support for investment in sustainable land management technologies as an imperative and integral element of farm activities.

3.8 Biodiversity and Conservation Agriculture

Kumaraswamy, S. and Kunte, K. (2013) Their study focusses on the major challenge to achieve food security and meet the demands of the ever-growing human populations while simultaneously conserving biodiversity, providing critical ecosystem services and maintaining rural livelihoods. Agro-ecology is proclaimed to be deteriorating due to eroding integrity of connected ecological mosaics and vulnerability to climate change. The authors suggested to implement offset mechanisms, which are compensatory and balancing approaches to restore the ecological health and function of an ecosystem to mitigate this devastating situation and integrate biodiversity with agricultural landscapes. Offset mechanisms can also be used to bridge the link between complementary conservation initiatives to design local to region-specific and conservation-oriented farming systems.

Khan, Zeyaur; Midega, Charles; Pittchar, Jimmy; Pickett, John and Bruce, Toby (2011)studied about the Push–pull technology which is based on a novel cropping system developed by the International Centre of Insect Physiology and Ecology, Rothamsted Research (UK) and national partners for integrated pest, weed and soil management in cereal–livestock farming systems. Stemborers are attracted to Napier grass (Pennisetum purpureum), a trap plant (pull), and are repelled from the main cereal crop using a repellent legume intercrop (push), desmodium (Desmodium spp.). Desmodium root exudates effectively control the parasitic striga weed by causing abortive germination. Desmodium also improves soil fertility through nitrogen fixation, natural mulching, improved biomass and control of erosion.Both companion plants offer high worth animal fodder, facilitating milk production and diversifying farmers' financial gain sources. The technology is suitable to smallholder mixed cropping systems in Africa. It effectively addresses major production constraints, increases maize yields from below 1 to 3.5t/ha, and is economical because it is predicated on regionally obtainable plants, not costly external inputs.

Dumanski, J., R. Peiretti, J. Benetis, D. McGarry, and C. Pieri (2006)studied the relationships between Conservation agriculture and zero tillage. Conservation agriculture is an application of modern agricultural technologies to improve production while concurrently protecting and enhancing the land resources on which production depends. Application of CA promotes the construct of optimizing yields and profits whereas making certain provision of native and world environmental advantages and services. Zero tillage, along with other soil conservation practices, is the cornerstone of CA. About 47% of the 95 million ha of zero tillage is practiced in South America, 39% in North America, 9% in Australia, and 3.9% in Europe, Asia and Africa.

Mcneely, Jeffrey A. and Schroth, Go¨tz (2006) studied that the traditional, often complex agroforestry systems are more supportive of biodiversity than mono-crop systems, although even they are no substitutes for natural habitat on whose proximity they may often depend for high levels of wild biodiversity. The relationship between forests, agroforestry and wild biodiversity can be made most productive through applying adaptive management approaches that recognize local knowledge and practice and incorporate ongoing research and monitoring in order to feed information back into the management system, with farmers and local populations included as active participants. The contributions highlight the considerable potential of traditional agroforestry practices to support biodiversity conservation, but also show their limits.These embody the importance of ample areas of natural surroundings and of appropriate hunting regulations for maintaining high levels of biodiversity in agroforestry land use mosaics, as well as the critical role of markets for tree products and of a favourable policy environment for agroforestry land uses.

3.9 Long Term Prediction in Ecological Study

Lindenmayer, David B.; Likens, Gene E.; Andersen, Alan; Bowman, David; Bull, C. Michael; Burns, Emma; Dickman, Chris R.; Hoffmann, Ary A.; Keith, David A.; Liddell, Michael J.; Lowe, Andrew J.; Metcalfe, Daniel J.; Phinn, Stuart R.; Russell-Smith, Jeremy; Thurgate, Nikki and Wardle, Glenda M. (2012) suggested that the ecological research community needs to put higher priority on communicating the benefits of long-term ecological studies to resource managers, policy makers and the general public. Long-term research will be especially important for tackling large-scale emerging problems confronting humanity such as resource management for a rapidly increasing human population, mass species extinction, and climate change detection, mitigation and adaptation. The authors also discussed about the five key values of such studies which are: (1) quantifying ecological responses to drivers of ecosystem change; (2) understanding complex ecosystem processes that occur over prolonged periods; (3) providing core ecological data that may be used to develop theoretical ecological models and to parameterize and validate simulation models; (4) acting as platforms for collaborative studies, thus promoting multidisciplinary research; and (5) providing data and understanding at scales relevant to management, and hence critically supporting evidence-based policy, decision making and the management of ecosystems.

Doak, Daniel F.; Estes, James A.; Halpern, Benjamin S.; Jacob, Ute; Lindberg, David R.; Lovvorn, James; Monson, Daniel H.; Tinker, M. Timothy; Williams, Terrie M.; Wootton, J. Timothy; Carroll, Ian; Emmerson, Mark; Micheli, Fiorenza and Novak, Mark (2008) conducted a study on ecological surprises, substantial and unanticipated changes in the abundance of one or more species that result from previously unsuspected processes, are a common outcome of both experiments and observations in community and population ecology. The researchers classify surprises as dynamic-, pattern-, or intervention-based, and we speculate on the common processes that cause ecological systems to so often surprise us. Most of the work on this subject involves statistical aspects of data analysis and modeling, the frequency and nature of ecological surprises imply that uncertainty cannot be easily tamed through improved analytical procedures, and that prudent management of both exploited and conserved communities will require precautionary and adaptive management approaches.

Froyd, C.A. and Willis, K.J. (2008) conducted a study on late Quaternary environmental change holds great promise for addressing many emerging issues in biodiversity and conservation management in the up-coming decade. This paper reviews recent developments in the application of the analysis of late Quaternary environmental change to key environmental issues of biodiversity and conservation management and examines areas which could be strengthened in the future including: (i) determination of baselines and natural ecosystem variability; (ii) understanding ecological thresholds and resilience; (iii) climate change conservation strategies; (iv) biological invasions; and (v) conservation and culture.

Willis, Katherine J; Araújo, Miguel B; Bennett, Keith D; Figueroa-Rangel, Blanca; Froyd, Cynthia A and Myers, Norman (2007)conducted a study where they have evaluated how long-term records could and should be utilized in conservation policy and practice. Key conservation initiatives related to extinctions, identification of regions of greatest diversity/threat, climate change and biological invasions are addressed. The paper concludes that often inclusion of a long-term ecological perspective can provide a more scientifically defensible basis for conservation decisions than the one based only on contemporary records. The pivotal issue of this paper is not whether long-term records are of interest to conservation biologists, but how they can actually be utilized in conservation practice and policy.

Willis, K. J. and Birks, H. J. B. (2006) conducted a study on ecosystems change in response to factors such as climate variability, invasions, and wildfires. Most records used to assess such change are based on short-term ecological data or satellite imagery spanning only a few decades. In several instances it's not possible to disentangle natural variability from alternative, probably important trends in these records, partly due to their short continuance.Summary of recent studies that show how paleoecological records can be used to provide a longer temporal perspective to address specific conservation issues relating to biological invasions, wildfires, climate change, and determination of natural variability has been mentioned in the paper. The use of such records will scale back a lot of the uncertainty surrounding the question of what's "natural" and thereby begin to produce vital guidance for long management and conservation.

Giangrande, Adriana (2003) This paper highlights the poor esteem within which taxonomy, as autonomous science, is command and also the relative implications of this for conservation biology. In recent times, taxonomy at the species level has tended to be neglected not simply among ecological researches, however conjointly within the identification and justification for the choice of the marine protected areas (MPA). several species seem to possess overlapping niches, and intrinsically it may be argued that it's not essential for all species to be present. In distinction the crucial role of keystone species has been embraced in conservation biology as a tool to assist highlight species requiring priority for cover.

Lindenmayer, D.B. (1999) conducted a study on the validity and use of the indicator species conception, the look of work impact studies, the necessity for long-term monitoring programs and the way they may be designed, and, trade-offs between conservation ways and economic costs are topics crucial to the long run direction of biological conservation in managed forests. The indicator species conception will build a very important contribution to biodiversity conservation owing to the impossibility of monitoring all taxa in species-rich forest environments. However, the concept has yet to be rigorously tested by validating relationships between an indicator species and entities for which it is hypothesized to be indicative. There can be serious negative consequences if the indicator species concept is incorrectly applied or inappropriate species are selected as indicators. Treating forestry activities as an experiment and overlaying well-designed monitoring programs on such disturbance regimes is one useful way to accumulate key information on the effects of logging on biodiversity and how to mitigate such impacts.

3.10 Entrepreneurship and Conservation Agriculture

Singh, SP; Naresh, RK; Panwar, AS; Vivek; Mahajan, NC; Kumar, Sudhir; Tomar, SK; Mrunalini, Kancheti and Prasad, KS Krishna (2019)discussed about thewider adoption of CA technologies requires concerted effort of all the stakeholders in the expanded partnership and participatory approaches in which farmers' experiments and provides rapid feed-back. This would should be upheld by institutional changes that advance information sharing, adaptability and decentralized dynamic for fast reception of innovations to keep up with creation and usefulness, expanded food security and business of the farmers.

Kassam, A.; Friedrich, T. and Derpsch, R. (2018)conducted a study on Conservation Agriculture (CA) which comprises the practical application of three interlinked principles, namely: no or minimum mechanical soil disturbance, biomass mulch soil cover and crop species diversification, in conjunction with other complementary good agricultural practices of integrated crop and production management. The need and the main reasons for adoption and the global spread of conservation agriculture are mentioned in this study.

Kimmel, Courtney E. and Hull, R. Bruce (2012) studied about the changing social and economic drivers of land use require a more integrated approach to land conservation that addresses both the land and the socio-economic context of land use. This research examines ecological entrepreneurship, discussed by Marsden and Smith (2005), as an integrated conservation strategy that targets environmental as well as economic goals. Specifically, the authors explored the roles and functions that must be provided by Ecological Entrepreneurship Support Networks (EESNs). In particular, they looked at land trusts engaging in these efforts. Through a series of case studies, they examined the roles and functions, as well as motivations and challenges faced by land trusts involved in EESNs. The exercises learned through these contextual analyses are expected to educate more extensive practical advancement endeavours.

Lambooy, Tineke and Levashova, Yulia (2011) found that private companies and investors can profit from the enhancement of nature in general and from specific investments allocated to improve biodiversity and ecosystem services (BES). This article demonstrates that new markets and business models are developing which are based on BES, thereby offering investment opportunities and contributing to nature conservation at the same time. This article gives an analysis of the various business models and the factors that support a proper functioning thereof, including the dependence on public regulation and the necessity to collaborate with local communities, authorities, NGOs or other stakeholders. In the analysis, barriers for attracting mainstream capital from institutional investors into 'BES businesses' are identified and addressed.

Seidl, Irmi; Schelske, Oliver and Joshi, Jasmin (2003) their study discussed the protection of biodiversity by means of market-based activities in the food sector and the preconditions and context that provide fertile ground. The variables investigated are the regional production factors involved, the entrepreneurship demonstrated, the kind of niche market developed, and the various kinds of support provided. A case study of a Swiss market activity designed to restore and protect biodiversity is presented; it concerns the cultivation of a traditionally grown wheat species combined with measures to protect wildlife and flora.

3.11 Barrier to Conservation Agriculture

Roesch-McNally, Gabrielle E.; Basche, Andrea D.; Arbuckle, J.G.; Tyndall, John C.; Miguez, Fernando E.; Bowman, Troy and Clay, Rebecca (2017) studied thatdespite the numerous environmental and production benefits to using cover crops, their adoption across agriculturally intensive regions such as Iowa is quite low. To better understand this knowledge gap, the authors conducted four focus groups across Iowa to examine research questions concerning field-level and structural barriers to cover crop adoption and strategies that farmers have used to overcome barriers in order to integrate cover crops in their management system. The analysis of these conversations uncovers a complex set of field-level and structural factors that farmers must navigate to make cover crop use effective in their operation.The investigation gives proof that diminishing structural boundaries to adoption might be important to build the utilization of this conservation practice to lessen ecological effects related with serious agricultural production.

Mwase, Weston; Sefasi, Abel; Njoloma, Joyce; Nyoka, Betserai I.; Manduwa, Daniel and Nyaika, Jacinta (2015) This paper looks at the elements that influence adoption of agroforestry and integrated soil fertility management (ISFM) practices in Southern Africa. Agroforestry practices, particularly evergreen agriculture and conservation agriculture with trees have arisen as sustainable measures of addressing to land degradation and loss of soil fertility. In spite of the fact that agroforestry is known to be beneficial to farmers and the environment, its selection rate falls a long way behind the projected objectives. The current investigation evaluated a few publications on selection of agroforestry in Southern Africa and complemented the review with household and key source meetings to get proof from farmers and promoters of the advancements on the variables influencing adoption.

Perry-Hill, R. and Prokopy, L.S. (2014) This examination analyses the distinctions in awareness, attitudes, constraints, and behaviours of agricultural, small agricultural, and rural residential landowners. The creators of the study led a mail survey of a stratified random sample of rural landowners owning two acres or more in the Little Calumet-Galien Watershed in north-western India. They found that small agricultural and rural residential landowners have more uplifting perspectives, yet are less acquainted with water quality issues and protection associations. Additionally, a small agricultural or rural residential landowner is related with more prominent readiness to attempt conservation practices. For all landowners, perceived cost was the greatest obstruction to conservation practice adoption. In view of these outcomes, the creators proposed that to be fruitful, conservation programs targeted at small agricultural and rural residential landowners should address nonpoint source pollution and conservation practice awareness, as well as the cost of training execution.

Pittelkow, Cameron M.; Liang, Xinqiang; Linquist, Bruce A.; Groenigen, Kees Jan van; Lee, Juhwan; Lundy, Mark E.; Gestel, Natasja van; Six, Johan; Venterea, Rodney T. and Kessel, Chris van (2014)led an investigation on, as far as possible and possibilities of the acts of Conservation farming which addresses a bunch of three crop management principles that has received solid global help to take care of the developing and really requesting populace with decreased outside inputs and negligible effects on the environment. A worldwide meta-analysis utilizing 5,463 matched yield observations from 610 investigations to analyze no-till, the original and central concept of conservation agriculture, with conventional tillage practices across 48 crops and 63 countries was performed. Results from this examination shows that no-till lessens yields, yet this reaction is variable and under specific conditions no-till can create same or more prominent yields than conventional tillage. However, any expansion of conservation agriculture should be done with caution, as execution of the other two standards is frequently difficult in asset poor and weak smallholder cultivating frameworks, thereby improving the probability of yield losses instead of gains. The investigation from this examination shows that the potential contribution of no-till to the sustainable intensification of agriculture is more restricted than frequently accepted.

Greiner, Romy and Gregg, Daniel (2011)Their investigation presents the results of an analysis of a pooled dataset, which was built by consolidating information from three separate studies of graziers inside the tropical savannas in Australia. The reviews had extensively various goals yet all were completed within 18 months of each other with an expectation of examining the connections between graziers' motivations, the significance of obstructions and the adequacy of incentives with reference to the adoption of conservation practices. The paper exhibits the benefit of planning research in a way that empowers esteem added investigation so that, on one hand, the consequences of contextual analyses can be approved and summed up and, on the other hand, local contrasts can be elicited and adequately considered to be in policy design and program execution.

Knowler, Duncan &Bradshaw,Ben (2007) conducted a study on farmers' adoption of conservation agriculture.This paper reviews and synthesizes the past research in order to identify those independent variables that regularly explain adoption, and thereby facilitate policy prescriptions to augment adoption around the world. The study concludes that efforts to promote conservation agriculture will have to be tailored to reflect the particular conditions of individual locales.

3.12 Extraction of Groundwater and its Impact on Ecological Resilience

Arfanuzzaman, Md. & Atiq Rahman, A.(2017)studied about the necessity of Sustainable water demand management (SWDM) is immensely higher in the rapidly urbanized mega cities of the world where groundwater depletion and water deficit are taking place perilously. This paper centers around the current state of water demand, supply, system loss, pricing strategy, groundwater level, and per capita water utilization of Dhaka city, Bangladesh. The study founds population growth has a large influence on water demand to rise and demand of water is not responsive to the existing pricing rule adopted by Dhaka Water Supply and Sewerage Authority (DWASA). The study proposes an integrated SWDM approach, which incorporates optimum pricing, ground and surface water regulation, water conservation, sustainable water consumption and less water foot print to ease groundwater depletion.

Grönwall, Jenny& OduroKwarteng, Sampson(2017)found that water insecurity is a developing concern worldwide, particularly for developing countries, where a scope of variables including urbanization are squeezing water provisioning frameworks. The role of groundwater and springs in buffering the impacts of environment fluctuation is progressively recognized, yet it must be completely acknowledged with a more powerful understanding of groundwater as a resource, and how utilization of it and reliance on it contrast. Conjunctive use, managed aquifer recharge, and suitable treatment measures are vital to make groundwater an essential asset on the metropolitan plan.

Katic, Pamela & Grafton, R. Quentin (2011) Their examination assesses the trade-offs among resilience and economic payoffs in terms of groundwater extraction where there is a danger of an irreversible and cataclysmic occasion. The researchers in their investigation fostered a dynamic and spatial model that joins a stochastic recharge process and the danger of an irreversible cataclysmic occasion (like saltwater intrusion) that emerges when water powered heads fall under a given threshold. The outcomes show that assuming the limit is dubious, controlling both the rate and profundity of extraction can create a higher monetary return and a lower likelihood of crossing the threshold than just controlling the rate of extraction. The model and discoveries give an applied system to comprehend and to measure where there may be 'shared benefit' results, and trade-offs between monetary adjustments and resilience in terms of groundwater extraction.

3.. Soil Erosion and its Impact on Ecological Resilience

Flores, Bernardo M; Staal, Arie; C, Catarina; Jakovac; Hirota, Marina; Holmgren, Milena & Oliveira, Rafael S. (2019)led an investigation on tropical woods which are threatened by increasing natural and anthropogenic aggravation systems. In view of a broad writing survey, the specialists proposed an applied model in which soil disintegration builds up unsettling influence impacts on tropical woods, diminishing their resilience with time and improving their probability of being trapped in an alternative vegetation that is vulnerable to erosion.

Kosmas, Costas; Karamesouti, Mina; Kounalaki, Kate; Detsis, Vassilis; Vassiliou, Penny; Salvati, Luca (2016) This study explores the socio-environmental framework dominated by pasture land in Asteroussia Mountains (Crete, Greece) between1950 and 2010 with the intend to distinguish changes in the most significant framework's credits, as a commitment to the investigation of land debasement according to the flexibility viewpoint. The grazing system was sustainable in the previous time span being agreeably dealt with tolerably low soil disintegration rates in spite of antagonistic soil, geographical and environment conditions.

Teague, W.R.; Apfelbaum, S.; Lal, R.; Kreuter, U.P.; Rowntree, J.; Davies, C.A.; Conser, R.; Rasmussen, M.; Hatfield, J.; Wang, T.; Wang, F.; and Byck, P. (2016)suggested that with suitable regenerative crop and grazing management, ruminants diminish generally GHG outflows, yet in addition facilitate provision of essential ecosystem services, increment soil carbon (C) sequestration, and decrease ecological harm. The analysts tried the theory by looking at biophysical impacts and the extent of all GHG outflows from key agrarian creation exercises, including examinations of arable-and peaceful based agroecosystems. The evaluation from the examination shows that internationally, GHG outflows from domestic ruminants address 11.6% (1.58 Gt C y–1) of absolute anthropogenic discharges, while cropping and soil-associated emissions contribute 13.7% (1.86 Gt C y–1). The paper infers that to guarantee long-term sustainability and ecological resilience of agroecosystems, agricultural production ought to be directed by arrangements and regenerative administration conventions that incorporate ruminant grazing.

Rescia, Alejandro J.; Willaarts, Bárbara A.; Schmitz, María F.; Aguilera,Pedro A. (2010)led an examination on the progressions in land use and management in the last decades which have prompted the corruption of numerous social scenes in European country regions. Such corruption has had pertinent ramifications for nearby populaces, landscape functionality, and the maintenance of ecosystem services. Successful administration of cultural landscapes, perceived as social ecological systems, ought to incorporate both, the social and ecological components. This administration ought to be versatile and include the dynamic interest of nearby populace.

Cumming, G. S.; Barnes, G.; Perz, S.; Schmink, M.; Sieving, K. E.; Southworth, J.; Binford, M.; Holt, R. D.; Stickler, C. and Holt, T. Van. (2005) studied about the idea of social-ecological resilience holds guarantee for interdisciplinary amalgamations. Resilience is a complex idea that at this point has not been straightforwardly operationalized, especially in systems for which our obliviousness is such that detailed, parameter-rich simulation models are hard to create. The analysts introduced an exploratory structure as a stage towards the operationalization of resilience for experimental investigations. They have equated resilience with the capacity of a system to keep up with its identity, where system identity is characterized as a property of key segments and connections (organizations) and their progression through space and time. This methodology will likewise yield experiences into the instruments of progress and the expected results of various approach and the management decisions, giving a degree of choice help for each contextual analysis region.

Lal,R. (1997)focused on global soil degradation, its extent and agronomic impact, which can only be resolved through understanding of the processes and factors leading to establishment of the cause-effect relationships for major soils, ecoregions, and land uses. Efficient assessment through long haul experimentation is required for setting up quantitative measures of (i) soil quality in relation to specific functions; (ii) soil degradation in relation to critical limits of key soil properties and processes; and (iii) soil resilience in relation to the ease of restoration through judicious management and discriminate use of essential input. There is a need to create and normalize strategies for estimating soil resilience.

3.14 Biodiversity and its Impact on Ecological Resilience

Oliver, Tom H.; Heard, Matthew S.; Isaac, Nick J.B.; Roy, David B.; Procter, Deborah; Eigenbrod, Felix; Freckleton, Rob; Hector, Andy; Orme, C. David L.; Petchey, Owen L.; Proença, Vânia; Raffaelli, David; Suttle, K. Blake; Mace, Georgina M.; Martín-López, Berta; Woodcock, Ben A. and Bullock,James M. (2015)highlighted on the mechanisms by which biodiversity, at different hierarchical scales, that influences the resilience of ecosystem functions. The scientists focused on resilience as opposed to transient conveyance of ecosystem functions and services, and the thought of specific underpinning mechanisms, which has assisted with joining the research areas of biodiversity– ecosystem function and ecological resilience and finally aid the development of evidence-based yet flexible ecosystem management.

Winn, Monika I. and Pogutz,Stefano (2013) This article plans to energize investigation into how organizations can deal with their relationship with the natural habitat so as not to annihilate the very life-supporting establishments provided by the nature. Spanning information areas, the article presents key ideas from ecology and social ecology to organization and management studies-ecosystems, biodiversity, ecosystem services, and ecological resilience. The researchers represent these ideas with progresses in ecosystems management and conclude with ideas for future examination in sustainability management, organization theory, and strategic management.

Jackson, Louise; Noordwijk, Meine van; Bengtsson, Janne; Foster, William; Lipper, Leslie; Pulleman, Mirjam; Said, Mohammed; Snaddon, Jake and Vodouhe, Raymond (2010) conducted a study about the quick changes in land use, food systems, and livelihoods require social-ecological systems that keep various choices open and get ready for future unpredictability. Sustainagility alludes to the properties and resources of a system that sustain the ability (agility) of agents to adapt and address their issues in new ways. In contrast, sustainability tends to invoke persistence along current trajectories, and the resilience to get back to current baselines. With three models, the utilization and conservation of agrobiodiversity is investigated along temporal, spatial, and human institutional scales for its role in sustainagility: first, farmers' seed systems; second, complex pollination systems; and third, wildlife conservation in agricultural regions with high destitution. Incentives are necessary if agrobiodiversity is to give advantages to people in the future.

Di Falco, Salvatore and Chavas, Jean-Paul (2008) Their study investigates the powerful impacts of rainfall shocks on agroecosystems productivity. The investigation estimates a panel data model of cereal production in southern Italy. It reports the adverse impacts of decrease in rainfall on the agroecosystem productivity both in the short run and long run. It researches how expanding the degree of spatial crop diversity can moderate this adverse consequence. The experimental proof shows how higher diversity upholds resilience and keeps up with the system productivity under challenging climatic conditions.

Moller, Henrik; J. MacLeod, Catriona; Haggerty, Julia; Rosin, Chris; Blackwell, Grant; Perley, Chris; Meadows, Sarah; Weller, Florian & Gradwohl, Markus (2008) led an examination where an obvious proof has been shown that agricultural intensification has degraded amphibian biodiversity, however there is a basic absence of research and monitoring of robust indicators of terrestrial biodiversity in New Zealand production landscapes. The examination gathers a summed up probability that intensification has likewise decreased terrestrial biodiversity and agro-ecosystem resilience. This survey closes with a crucial statement: a shift of emphasis is now desperately required inside New Zealand land-use management, research and policy. There is an improper progressing and historical emphasis on allotment models that see preservation and extractive utilization of regular assets as fundamentally unrelated alternate extremes that must be sufficiently pursued in different places. They should be supplanted by integrative models of conservation and use.

Fischer, Joern; Lindenmayer, David B. and Manning,Adrian D. (2006) suggested ten guiding principles to assist in maintaining biodiversity, ecosystem function, and resilience in production landscapes. Landscapes should incorporate fundamentally trademark patches of local vegetation, corridors and venturing stones between them, a structurally complex matrix, and supports around delicate regions. Management ought to maintain a variety of species inside and across functional groups. Exceptionally engaged management activities might be needed to maintain keystone species and threatened species, and to control invasive species. These guiding principles provide a scientifically defensible beginning stage for the combination of conservation and production, which is immediately needed from both an ecological and a long-term financial point of view.

Hughes, Terence P.; Bellwood, David R.; Folke, Carl; Steneck, Robert S. and Wilson,James (2005)observed that the vital component in social-ecological systems resilience-based management is the recognition of the linkages between the environment and people. Natural resources are too valuable, economically, culturally and aesthetically, to be wasted. Besides, re-establishing marine and terrestrial ecosystems after they have degraded is substantially more troublesome than maintaining them in good condition. New paradigms, perspectives, policies and governance systems are desperately expected to safeguard ecological systems for the development of the society and future generations. The investigation featured the rise of an intricate frameworks approach for sustaining and repairing marine ecosystems, connecting ecological resilience to governance structures, financial matters and society.

Perrings, Charles and Walker,Brian (1997)directed an investigation on the loss of resilience in systems portrayed by various equilibria which is indicated by an intermittent change in the state of the system, or the transition from one locally stable state corresponding to a specific blend of species to another state corresponding to an alternate blend of species. The resilience of the system may be affected by the management regime through its impact on the response to unsettling influences. The paper shows that ideal management of an event driven system should be sensitive to the chances made by such events.

3.15 Awareness, Knowledge and Action for the Restoration of Ecosystems

Su, Yang; Gabrielle, Benoit and Makowski, David (2021)their study reveals that no tillage (NT) is often presented as a means to grow crops with positive environmental externalities, such as enhanced carbon sequestration, improved soil quality, reduced soil erosion, and increased biodiversity. Notwithstanding, regardless of whether NT frameworks are pretty much as useful as those depending on traditional culturing (CT) is a disputable issue, full by a high fluctuation over the long run and space.The study expands existing datasets to include the results of the most recent field experiments, and produce a global dataset comparing the crop yields obtained under CT and NT systems. This dataset can help to gain insight into the main drivers explaining the variability of the productivity of NT and the consequence of its adoption on crop yields.

Stuart Chapin III, F.; Sommerkorn, Martin; Robards, Martin D.; Hillmer-Pegram, Kevin (2015) their study presents a stewardship framework for a more practical direction for arctic species, ecosystems, and peoples—to some extent expanding on promising existing endeavors, and somewhere else supporting new segments. The framework more thoroughly incorporates social and ecological elements of conservation across scales and anticipates and shapes multi-scale changes. Stewardship consequently gives an integrated framework for conservation practice, including useful direction for society to further develop conditions for conservation in a quickly evolving world.

Ballard, Heidi L. & Jill M. Belsky,(2010)led an investigation on Participatory action research (PAR) which is a way to deal with research that its supporters claim can foster new information, learning, and action to support positive social and environmental change through reorienting the standard cycle of information creation. In this paper, the analysts inspected a PAR venture to decide how PAR fostered environmental learning and, in turn, how the learning influenced resilience. They tracked down that the PAR approach did surely create environmental learning, characterized here as ecological literacy, civic literacy, values awareness and self-efficacy and added to resiliency through advancing more diversity, memory, redundancy, and adaptive capacity. The investigation inferred that the PAR approach is an important tool for environmental learning however the degree to which learning can really promote system change and greater resilience should likewise be perceived inside the basic setting, particularly political realities.

King,Christine A. (2008)conducted a study on current research which explore a range of agri-ecological systems and include systems such as Organic Agriculture, Biodynamics, Community Supported Agriculture (CSA), Permaculture, Farmers Markets and Community Gardens and how they contribute to agri-ecological and community resilience. The study reviewed these alternative approaches to food production in relation to key concepts from ecological systems thinking, such as ecological resilience, biodiversity and holism. The study concluded by comparing ecological systems models to agri-ecological systems, and suggests how ecological systems theories and concepts might contribute to thinking about the future of community-based agri-ecological resilience.

Fiksel, Joseph (2006) emphasized on a comprehensive system approach which is necessary for effective decision making with regard to global sustainability, since industrial, social, and ecological systems are firmly connected. There is a pressing requirement for a superior comprehension of the dynamic, adaptive behaviour of complex systems and their resilience despite disturbances, perceiving that consistent state sustainability models are simplistic. Various exploration bunches are utilizing dynamic modelling techniques, including bio complexities, system dynamics, and thermodynamic analysis, to research the effects on ecological and human systems of significant moves, for example, climate change and the related policy and technology responses. These methods can yield something like a partial understanding of dynamic system behaviour, empowering a more coordinated way to deal with systems analysis, beneficial intervention, and improvement of resilience. Recommendations are provided for continued research to accomplish progress in the dynamic modeling and sustainable management of complex systems.

Folke, Carl; Carpenter, Steve; Elmqvist, Thomas; Gunderson, Lance; Holling, C. S. and Walker, Brian (2002) summarized a report prepared on behalf of the Environmental Advisory Council to the Swedish Government, as input to the process of the World Summit on Sustainable Development (WSSD) in Johannesburg, South Africa in 26 August 4 September 2002. We use the concept of resilience—the capacity to buffer change, learn and develop— as a framework for understanding how to sustain and enhance adaptive capacity in a complex world of rapid transformations. Two useful tools for resilience-building in social-ecological systems are structured scenarios and active adaptive management. These tools require and work with a social setting with adaptable and open organizations and multi-level governance systems that allow learning and increment adaptive capacity without abandoning future improvement choices.

Olsson, Per and Folke, Carl (2001) Their study focused on the local management of crayfish, a common-pool resource, and also address the way crayfish management is linked to institutions at different levels of Swedish society. Methods from the social sciences were used for information gathering, and the results were analyzed within the framework of ecosystem management. The researchers found that the practices of local fishing association resemble an ecosystem approach to crayfish management. The results from the study indicated that local users have substantial knowledge of resource and ecosystem dynamics from the level of the individual crayfish to that of the watershed, as reflected in a variety of interrelated management practices embedded in and influenced by institutions at several levels. The researchers also suggested that this policy of monitoring at several levels simultaneously, together with the interpretation of a bundle of indicators and associated management responses, enhances the possibility of building ecological resilience into the watershed.

4

Methods and Approaches of Study

The deliberation on the methodology has been made to understand the concept, methods and techniques which are utilized to design the study, collection of information, analysis of data and interpretation of the findings for revelation of truths and formulation of theories. These chapter deals with the method and a procedure used in the study consisting eight main parts

A. Locale of research.

B. Pilot study.

C. Sampling Design.

D. Item analysis and experts rating for devising scoring technique on dependent variables.

E. Empirical measurement of the variables.

F. Preparation of interview Schedule.

G. Pre testing of Interview Schedule.

H. Techniques of Data collection.

I. Statistical Tools used for Analysis of Data.

4.1 Locale of Research

Kastodanga and Dalilpur village of Haringhata block of Nadia district in West Bengal was randomly selected for the study. The area has been selected for the study because of – a) There is sample scope for collecting relevant data for the present study, b) Acquaintance with the local people as well as local language, c) The concern area was easily accessible to the researcher, d) The closer familiarities of the student researchers with the area, people, officials and local dialects.

4.2 Pilot Study

Before taking up actual field work a pilot study was conducted to understand the area, its people, institution, communication and extension system and the knowledge, perception and attitude of the people towards climate changes concept.

4.3 Sampling Design

Purposive as well as simple random sampling techniques were adopted for the study.

Table 4.1 Sampling Techniques and Sampling Design

Sl. No.	Level	Approach	Remarks
1.	State-West Bengal	Purposive	High intensity agriculture has already depleted the agro-ecosystem of the state and district so it needs a research insight in agro-ecosystem study.
2.	District-Nadia	Purposive	
3.	Block-Haringhata	Purposive	High intensity agriculture, rice and vegetable based farming, heavy depletion of ground water, decline trend of productivity, livestock count and organic carbon. It necessitates the need for Conservation Agriculture.
4.	Villages-Kastodanga and Dalilpur	Purposive	Ease of accessibility
5.	50 respondents from both villages	Random and Non-random	It depends on Covid-19 situation, socio-political situation and level of responsiveness from the farmers. Collection of data from 50 respondents is the target.

4.4 Item Analysis and Experts Rating for Devising Scoring Technique on Dependent Variables

Since the topic of the research has been so new and unique, the present researcher and as per guidance of the supervisor has generated a score of marker questions to be rated on ten point scales by the selected experts comprising of eminent scientist in the concerned areas and extension practitioner widely exposed to the domain. The steps were as follows

a) Random selection of item.

b) Screening and scrutinizing of the items based on applicability and relevancy.

c) Finalization of marker item / statements to be rated on ten-point scale.

d) Selection of the experts and seeking their consents.

e) Sending the item to the selected experts and collecting their feedback (80 percent of the feedback from the experts were available).

f) Respective item values were summated.

g) 50 percent of the items were retained and rests were discarded.

h) Through before after test on 20 selected respondents, the r value was calculated and it was found significant.

i) The final marker statement has thus been finally selected and applied for data collection.

4.5 Empirical Measurement of the Variables

After reviewing various literatures related to the field of study and consultation with respected chairman Advisory Committee and other experts, a list of variables was prepared.

Table 4.2: List of Dependent variables

Sl. No.	Dependent Variables	Notations
1.	Application of organic manure	y_1
2.	Rice stubble height left in the field	y_2
3.	Crop residue left in field	y_3
4.	Climate resilient technology	y_4
5.	Soil conservation perception	y_5
6.	Water conservation perception	y_6
7.	Biodiversity conservation perception	y_7
8.	Soil stewardship perception	y_8
9.	Water stewardship perception	y_9
10.	Biodiversity stewardship perception	y_{10}

Application of organic manure (y_1)

Organic manure has been perceived as organic source of plant nutrient called manure either produced at domestic level or produced from elsewhere. It is measured in terms of kg/bigha/year applied in the field.

Rice stubble height left in the Field (y_2)

After harvesting some part of the paddy straw, the vegetative plant parts are left in the field. These are called stubbles and important source organic manures. The higher is the length of stubble, left, the more is the volume of the organic manure. Here it has been measured in terms of inches.

Crop residue left in the field (y_3)

Crop residue means the unused or uneconomic parts of the plants (Vegetables, pulsed etc.). These are the by product and also used as important source of organic manure. These are measured in terms of kg.

Climate resilient technology (y_4)

Climate resilient technology here has been conceived in the form of those approaches or technologies, modern or traditional followed to increase the copping or adapting ability of existing form against the brunt of global warming and climate change. Here it has been measured by scoring different measures taken up by farmers of the study area to cope up with the changing climate and environment.

Score 1= Not cultivating crop which are vulnerable prevailing climate and pest and disease infestation.

Score 2= Changes in the water management practices.

Score3= Cultivating vegetables or oilseed instead of boro rice.

Soil conservation perception (y_5)

Soil conservation perception has been measured in terms of selected item and these items are-

Soil Conservation Item	Score out of 10
1. Soil conservation helps protection of fertility of soil, productivity of crop thereafter.	
2. Soil conservation also helps in protection of microorganisms as well.	
3. We have enough of traditional knowledge on soil conservation, but needs proper technical support.	

The items were finally selected from the expert's rating on 10-point scale. Initially 6 items were selected and based on relevancy level 3 items were finally selected. This is another form of summated rating scale (Likert scale).

Water conservation perception (y_6)

Water conservation perception has been measured in terms of selected item and these items are-

Water Conservation Item	Score out of 10
1. Water conservation helps lifesaving irrigation during critical drought period.	
2. Destruction of natural water bodies amounts to loss of plant and fish biodiversity.	
3. Water conservation must focus on judicious use of groundwater, of more on rainwater harvesting	

The items were finally selected from the expert's rating on 10-point scale. Initially 6 items were selected and based on relevancy level 3 items were finally selected. This is another form of summated rating scale (Likert scale).

Biodiversity conservation perception (y_7)

Biodiversity conservation perception has been measured in terms of selected item and these items are-

Biodiversity Conservation Item	Score out of 10
1. Crop biodiversity conservation helps stability in production, better risk management and effective use of natural resources like land, water etc.	
2. Crop biodiversity conservation also helps protection of target crops from disease and pest attack.	
3. We have enough of traditional knowledge on crop biodiversity conservation, but needs proper technical support.	

The items were finally selected from the expert's rating on 10-point scale. Initially 6 items were selected and based on relevancy level 3 items were finally selected. This is another form of summated rating scale (Likert scale).

Soil stewardship perception (y_8)

Soil stewardship perception has been measured in terms of selected item and these items are-

Soil Stewardship Item	Score out of 10
1. When soil is made productive, it clearly contributes to better farm economy.	
2. For sustainable agriculture, productive function of soil needs resilience.	
3. Soil Stewardship is close to livelihood and economy wherein conservation is closure to natural resource and its ecology.	

The items were finally selected from the expert's rating on 10-point scale. Initially 6 items were selected and based on relevancy level 3 items were finally selected. This is another form of summated rating scale (Likert scale).

Water stewardship perception (y_9)

Water stewardship perception has been measured in terms of selected item and these items are-

Water Stewardship Item	Score out of 10
1. Acknowledge indigenous knowledge, customary laws and informal institution in water management. 2. Conjunctive use of water increases both economic and ecological proficiency. 3. Water Stewardship helps attaining better ecological productivity, economic growth and resilience.	

The items were finally selected from the expert's rating on 10-point scale. Initially 6 items were selected and based on relevancy level 3 items were finally selected. This is another form of summated rating scale (Likert scale).

Biodiversity stewardship perception (y_{10})

Biodiversity stewardship perception has been measured in terms of selected item and these items are-

Biodiversity Stewardship Item	Score out of 10
1. Crop biodiversity per unit area reduces risks and enhances profitability.	
2. Biodiversity should be conserved because it fosters ecological resilience as well as productivity.	
3. A crop mix has got certain benefits over monolithic agriculture because it integrates benefits and countervails negativity.	

The items were finally selected from the expert's rating on 10-point scale. Initially 6 items were selected and based on relevancy level 3 items were finally selected. This is another form of summated rating scale (Likert scale).

Table 4.3: List of Independent Variables

Sl. No.	Independent Variables	Notations
1.	Age	x_1
2.	Education	x_2
3.	Functional education	x_3
4.	Family size	x_4
5.	Size of land holding	x_5
6.	No. of land fragments	x_6
7.	Average size of land fragment	x_7
8.	Mean distance between two land fragments	x_8
9.	Size of homestead land	x_9
10.	No. of crops cultivated	x_{10}
11.	Total yield of crops cultivated	x_{11}
12.	Total marketed surplus of crops cultivated	x_{12}
13.	Total input cost	x_{13}
14.	On farm income	x_{14}
15.	Family expenditure/year	x_{15}
16.	Savings/year	x_{16}
17.	Land under irrigation	x_{17}
18.	Cropping intensity	x_{18}
19.	Total hours of irrigation given	x_{19}
20.	Livestock count	x_{20}
21.	Communication variable	x_{21}
22.	Consumption of coal, firewood, fuelwood	x_{22}
23.	Production of organic manure	x_{23}

Age (x_1)

In all societies, age is one of the most important determinants of social status and social role of the individual. In the present study, age of the respondent was measured on the basis of their chronological age at the time of investigation.

Education (x_2)

Education is instrumental is building personality structure and helps in changing one's behavior in social life. Education may be conceptualized as the amount of formal schooling literacy acquired by the respondent.

Functional Education (x_3)

The respondents were asked how many trainings they have participated or attended regarding the technical knowledge of cultivation and also post-harvest processing, value addition or any other activities from KVKs, SAUs etc. It has been measured on a score of ten-point scale-

Score 3- Informal education

Score 6- Attended two to five trainings

Score 9- Attended more than five trainings

Size of family members (x_4)

It is the number of family members of individual farmers.

Size of land holding (x_5)

The amount of land owned by a person is an important parameter to access the economic status of the person in the society. Here it has been measured by total area of land owned by the respondent in terms of bigha.

Number of fragment (x_6)

Land fragments here have been perceived in terms of total number of entitlements under the total size of holding. For example – if one ha of land has got three geographical distributions than the number of fragmentation will be three. Here it has been measured by counting the arithmetic count of pieces of the land (fragments) under the ownership of family.

Average size of fragment(x_7)

Average or mean size of total number of fragments were taken into account in terms of katha.

Mean distance between two fragments(x_8)

Mean distance between all fragments under the ownership of a family were measured in terms of metre.

Size of homestead land(x_9)

Amount of land acquired by the home building and surrounding. This total area was measured in katha.

No. of crops cultivated(x_{10})

It is total number of crops cultivated by the respondents throughout the cropping season. Here it has been measured by counting number of crops cultivated by the farms.

Total yield of crops cultivated(x_{11})

Total yield of all the crops (rice, vegetables, pulses etc.) throughout the cropping season were measured in kg for each family.

Total marketed surplus of crops cultivated(x_{12})

Total amount of produced sold to the market by each respondent were measured in kg.

Total input cost(x_{13})

Total input cost is the total expenditure for buying seeds, fertilizers, pesticides as well as the labour cost incurred by the farmers. Here it has been measured by adding all the input cost (i.e. the cost of seeds, fertilizers, pesticides and labour) for all the crops that the farmer cultivates in his land throughout the cropping season in terms of rupees.

On farm income(x_{14})

Total income by the respondent from farm in a year was measured in rupees.

Family expenditure(x_{15})

Total family expenditure of the farmer in a year was measured in terms of rupees.

Savings(x_{16})

Total savings of the farmer in a year was measured in terms of rupees.

Land under irrigation(x_{17})

Total area under the ownership of one family covered under irrigated agriculture was measured in bigha.

Cropping intensity(x_{18})

It refers to the proportion of acreage annually put under different crops to the total cropped area as expressed in percentage. The cropping intensity is calculated by

Cropping intensity (%) = (Gross Cropped Area/ Net Cropped Area) *100

Total hours of irrigation given(x_{19})

Total hours of irrigation given in each season to rice, vegetables and pulses and other crops cultivated by the respondent were measured in hours.

Livestock count(x_{20})

Arithmetic count cattle, poultry, goat possessed by each family were taken into account.

Communication variable(x_{21})

It refers to the use of information source(s) both from outside and within the community for getting agricultural information. The score assigned as followed:

TV/Radio (**A**)	Newspaper or any print media (**B**)	Dealers or peer farmer (**C**)	Experts from an Institution (**D**)	Calculation
A days in a week	**B** days in a week	**C** days in a week	**D** days in a week	**A***12+**B***12+**C***12+**D***12

Consumption of coal, firewood, fuelwood(x_{22})

Consumption of green sources of energy is a common habit in rural India. It reduces their cost of purchasing fuel mainly for culinary purpose from outside. Here it has been measured in terms of kg of coal, fuel wood, fire wood, bamboo, dried and uneconomic part of crop used in each year by the family.

Production of organic manure(x_{23})

Cow dung, poultry or goat litter, biogas slurry are an important domestic by product which subsequently improve the soil health and also crop quality and usher the possibilities of stewardship. Here it has been measured by taking into account the total amount of cow dung, poultry or goat litter, biogas slurry produced in a year and it was measured in kg.

4.6 Preparation of Interview Schedule

On the basis of finding pilot study a preliminary schedule was formed with the help of literature and by the assistance of Chairman of Advisory Committee. The interview schedule consisted of three major parts according to the specific objective of the study.

4.7 Pre Testing of Interview Schedule

It is the process of advance testing of the study design after the schedule has been prepared. The object of pretesting is to detect the discrepancies that have emerged and to remove them after necessary modification in the schedule. After conducting pretesting appropriate changes and modification of the interview schedule have been made. The individuals who responded in pretesting have been excluded in the final sample selected for the study.

4.8 Techniques of Field Data Collection

Data collection was done by the researcher herself from the farmers of Dalilpur and Kastodanga villages. Telephonic interview based on the interview schedule was conducted. It was depended on Covid-19 situation, socio-political situation and level of responsiveness from the farmers. The contact numbers of the respondent of Kastodanga village were collected from the Kastodanga Gram Samabay Krishi Unnayan Samiti Limited and contact numbers of the respondent of Dalilpur village were collected from few farmers of the village Dalilpur.

4.9 Statistical Tools used for Analysis of Data

The statistical methods for analysis and interpretation of raw data are-

a) Range.

b) Mean.

c) Standard deviation.

d) Coefficient of Variance.

e) Correlation of coefficient.

f) Multiple regression analysis

g) Stepwise regression analysis.

h) Path analysis.

i) Factor analysis.

j) Artificial neural network.

a) Range: The RANGE of a distribution is the difference between the smallest and largest of observations and gives us some idea of the amount of variability present (Panse and Sukhatme, 1967). The range is the most general measure of spread or scatters and may be computed to make a rough comparison of two or more groups for variability.In statistics, the range of a set of data is the difference between the largest and smallest values. It is expressed by maximum and minimum value of a set of data.

b) Mean: The mean is the arithmetic average and is the result obtained when the sum of the values of individuals in the data is divided by the number of individuals in the data (Panse and Sukhatme, 1967). Mean is the simplest and relatively stable measure of central tendency. The mean reflects and is affected by every score in the distribution. Thus, extreme scores affect the mean.

The Mean is computed by the formula

$$\overline{X} = \frac{\sum X}{N}$$

Where, $\overline{X}$ = mean of the scores

$\sum x$ = sum of individual scores

N = number of observations

When the data are expressed in a frequency distribution (grouped), the mean is calculated using the following formula

$$\overline{X} = \frac{\sum f.x}{N}$$

Where,

$\overline{X}$ = Mean of the observation.

f = Frequency of the class.

x = Mid value of the class.

N = Number of observations

c) Standard deviation: Standard deviation (SD) of a set of observation is the square root of the arithmetic mean of the squares of all deviations, the deviations being measured from the arithmetic mean of the distribution. It is commonly denoted by the symbol σ (Sigma). It is less affected by sampling error and is a more stable measure of dispersion. The standard deviation of the data grouped in the form of frequency distribution is computed by using the following formula-

$$\sigma = \sqrt{\frac{\sum f.d^2}{N}}$$

Where,

f= frequency of the class

d = deviation of the mid-value of the class from the population mean

N = total number of observations

d) Coefficient of Variance: A measure of variation which is independent of the unit of measurement is proved by the coefficient of variation. Being unit free, this is useful for comparison of variability between different populations. The coefficient of variation is standard deviation expressed as percentage of the mean. Coefficient of variation is measured by using the following formula —

$$CV = \frac{SD}{Mean} \times 100$$

e) Coefficient of correlation: When increase or decrease in one variety is accompanied by an increase or in the other variate, the two are said to be correlated and the phenomenon is known as CORRELATION. Correlation coefficient (r) is a measure of the relationship between two variables which are at the interval or ratio level of measurement and are linearly related. Pearson product-moment "r" is computed by the formula.

$$r_{xy} = \frac{N\sum XY - (\sum X)(\sum Y)}{\sqrt{\left[N\sum X^2 - (\sum X)^2\right]\left[N\sum Y^2 - (\sum Y)^2\right]}}$$

When,

X and Y = original scores in variables X and Y

N = number of paired scores

$\sum XY$ = each *X* multiplied by its corresponding *Y*, then summed

$\sum X$ = sum of *X* scores

$\sum X^2$ = each *X* squared, then summed = each *X* squared, then summed

$(\sum X)^2$ = sum of *X* scores, squared

$\sum Y$ = sum of *Y* scores = sum of *Y* scores

$\sum Y$ = sum of *Y* scores = each *Y* squared, then summed

$\sum Y$= sum of *Y* scores = sum of *Y* scores, squared

The range of correlation coefficient is between -1 to +1. This means that -1 is perfect negative correlation, +1 is perfect positive correlation. If the numbers of errors increase with increase in typing speed, it indicates positive correlation and if the numbers of correct words decrease with increase in typing speed, it is indicative of negative correlation.A correlation coefficient to be acceptable should be statistically significant. Otherwise, we say that no significant relationship exists between the variables.

f) Multiple regression analysis: A multiple correlation coefficient measures the combined relation between a dependent (consequent) and a series of independent (causal) variables. It can also be explained as the correlation between the observed values of the dependent variable and its estimated values from the independent variate values, estimated with the help of the multiple regression equation (Chandel,1970)

If **y** is the consequent variable and x_1, x_2 .... are the causal variables then the multiple regression equation is given by the following formulas –

$$y = a + b_1x_1 + b_2x_2....$$

Where, a is a constant and b_1, b_2, ... are the partial regression coefficients

The square root of the ratio of the regression sum of squares to the total sum of squares is known as multiple correlation coefficient and is denoted by ***R***. It is always positive and less than one. The square of the multiple correlation coefficient R^2 represents the fraction of the variation in **y**, accounted for by its joint association with the variates x_1, x_2 .... The coefficient is, therefore a measure of the joint association of all these variates with the dependent variate *y*, and tells us how much of the variation in *y* could be accounted for by reference to these variates (Panse and Sukhatme, 1967).

The beta coefficient is**the degree of change in the outcome variable for every 1-unit of change in the predictor or causal variable**. ... If the beta coefficient is positive, the interpretation is that for every 1-unit increase in the predictor variable, the outcome variable will increase by the beta coefficient value.

g) Stepwise regression analysis: Stepwise regression is a variation of multiple regression which provides a means of choosing independent variables that yield the best prediction possible with the fewest independent variables. It permits the user to solve a sequence of one or more multiple linear regression problems by stepwise application of the least square method. At each step in the analysis, a variable is added or removed which results in the greatest production in the error sum of squares (Burroughs Corporation, 1975)

According to **Draper and Smith** (1981), the method of stepwise multiple the analysis is to insert variables in turn until the regression equation is satisfactory. The order of insertion is determined by using the partial correlation coefficient as a measure of the importance of variables not yet in the equation. The programme, according to **Burroughs Corporation** (1975), first forms a correlation matrix, finds the best predictor (the independent variable having the highest correlation with the criterion variable) and performs a regression analysis with this predictor. Then. the second best predictor (independent variable having the second highest correlation with the criterion) is found and regression analysis using the multiple correlations of the two best predictors is performed, and so on. At any given step, the group of predictors being used is not necessarily the best group of that size (i.e. the particular group of independent variables does not necessarily have the highest multiple correlations with the criterion that any group of this size does).

Rather, this group contains the variables that have the highest individual correlation with the criterion. Significance of a variable that is being considered for entrance into the regression equation is measured by the F-statistic. If F is too small (less than F 'include'), the variable is not added to the regression equation. Include statement establishes the minimum value of the F-statistic required for the inclusion of a variable in the regression equation. In the example which follows, the F-value for inclusion was 0.01.

Significance of variables already in the regression equation may change as new variables are entered. This significance of the variables currently in the equation is also measured by the F-statistic. If F is too small (less than F 'delete'), the variable is not added to the equation. Delete establishes the value of the F statistic below which the variable is deleted from the regression equation. Here the F-value for deletion was 0.005. The 'tolerance' level specified is used as control of degeneracy. Degeneracy, when a variable entered into the equation, is a linear combination of variables present in the equation. Tolerance statement establishes the maximum value a pivoted element may attain while still allowing its associated variable to be brought into equation. A variable is not brought into the regression equation if its associated pivoted element is below the specified tolerance level, which was 0.001 in the present example.

h) Path Analysis: The objective of doing Path Analysis is to get a clear picture of the direct and indirect effects of the independent variables on the dependent variable. Variables, through which substantial indirect effects are channelled, are also found out.

Singh and Chaudhary (1977), defined path coefficient as the ratio of the standard deviation of the effect due to a given cause to the total standard deviation of the effect i.e. if **Y** is the effect and $\mathbf{x}_1$is the cause, the path coefficient for the path from cause $\mathbf{x}_1$ to the effect **Y** is

$$\sigma \, x_1 / \sigma \, \mathbf{y}$$

It is advisable to do path analysis with only those variables which have significant effects on the dependent variable. This may be done by restoring to multiple regression analysis and selecting those independent variables whose partial b value are significant. This shall enhance the clarity of the path analysis.

i) Factor Analysis: Factor analysis is a very useful and popular method of multivariate research technique, mostly used in social and behavioral sciences. It is a technique applicable when there is a systematic interdependence in finding out something more fundamental or latent which creates this commonality. For instance, we might have data, say, about an individual's income, education, occupation, and dwelling area and want to infer from these some factor (such as social class), which summarizes the commonality of all the said four variables.

Factor analysis seeks toresolve a large set of measured variables in terms of relatively few categories, known as factors. This technique allows the researcher to group variables into factors (based on correlation between variables), and the factors so derived may be treated as new variables (often termed as latent variables) and their value derived by summing the values of the original variables, which have been grouped into the factor. The meaning and name of such new variable is subjectively determined by the researcher (Kothari,1996).

Since the factors happen to be linear combinations of data, the coordinates of each observation or variable is measured to obtain what are *factor loadings*. Such factor loading represents the correlation between the particular variable and the factor, and are usually placed in a matrix of correlations between the variables and the factors. In the Factor Analysis the "Principle Component Analysis" was followed.

j) Artificial Neural Network: Artificial neural networks(ANNs), usually simply calledneural networks(NNs), are computing systems inspired by the biological neural networksthat constitute animal brains.

An artificial neural network (or simply neural network) consists of an input layer of neurons (or nodes, units), one or two (or even three) hidden layers of neurons, and a final layer of output neurons. Each connection is associated with a numeric number called weight. The output, $\mathbf{h}_i$, of neuron **i** in the hidden layer is,

$$h_1 = \sigma\left(\sum_{j=1}^{N} V_{ij} X_j + T_i^{hid}\right)$$

where, σ () = activation (or transfer) function,

N = number of input neurons,

V_{ij} =weights,

X_j = inputs to the input neurons,

T_i^{hid} = threshold terms of the hidden neurons.

The purpose of the activation function is, besides introducing nonlinearity into the neural network, to bound the value of the neuron so that the neural network is not paralyzed by divergent neurons. It has been shown that a neural network constructed the way above can approximate any computable function to an arbitrary precision. Numbers given to the input neurons are independent variables and those returned from the output neurons are dependent variables to the function being approximated by the neural network. Inputs to and outputs from a neural network can be binary (such as yes or no) or even symbols (green, red, ...) when data are appropriately encoded. This feature confers a wide range of applicability to neural networks.

The other essential ingredient of a neural network application, namely, *training*. Similar to human learning by examples, a neural network is trained by presenting to it a set of input data called *training set*. The desired outputs of the training data are known so that the aim of the training is to minimize, by adjusting the weights between the connected neurons, *an error function* which is normally the sum of the squared differences between the neural network outputs and the desired outputs.

If we are experimenting architectures of neural networks, an independent data set called *validation set* can be applied to the trained neural networks. The one which performs best is then picked up as the one of choice. After validation, yet another independent data set called *test set* is used to determine the performance level of the neural network which tells how confident we are when using the neural network. It should be understood that a neural network can never learn what is not present in the training set. The size of the training set has therefore to be large enough for the neural network to memorize the features/trends embedded in the training set. On the other hand, if too much unimportant details are contained in the training set, the neural network might waste its resources (weights) fitting the noise. A judicious selection and/or representation of the data are therefore critical to successful implementations of neural networks.

5

Case Study and Hard Evidences

The chapter deals with the results and discussion of the study, with a purpose to provide sufficient information so as to arrive at valid conclusion. Tables and models are commonly used in this chapter to support and supplement the textual materials. Discussion helps to interpret the results of the study in proper perspective and to relate them with other relevant studies.

The distribution pattern of variable, **Age (x_1)**, depicts that the minimum age of the respondents is 24 and maximum is 65. The mean age is 47.460 so most of the respondent belongs to middle age category with standard deviation 11.156. The coefficient of variance (CV) is 23.506 per cent which indicates that the distribution pattern of the variable is highly consistent.

The distribution pattern of variable, **Education (x_2)**, shows that the minimum standard of the respondents is 2 and the maximum is 10. The mean education is 5.600 with S.D of 2.969. The coefficient of variance is 53.022 per cent which indicates distribution of the variables is fairly consistent.

The distribution pattern of variable, **Functional education (x_3)**, shows that the minimum standard of the respondents is 3 and the maximum is 9. The mean of functional education is 4.560 with S.D of 2.032. The coefficient of variance is 44.561per cent which indicates distribution of the variables is fairly consistent.

The distribution pattern of variable, **Family size (x_4)**, depicts that the maximum family size of the respondents is 16 and the minimum is 2. The mean of size of the family is 4.780 with standard deviation of 2.306. The coefficient of variance is 48.244 per cent which indicates the distribution of variable is fairly consistent.

The distribution pattern of variable, **Size of land holding (x_5)**, depicts that the maximum size of holding is 22 bigha and the minimum is 0.70 bigha. The mean size of holding is 5.789 bigha with standard deviation of 4.615 bigha. The coefficient of variance is 79.723 per cent which indicates the distribution of variable is fairly consistent.

Table- 5.1: Descriptive Statistics of Independent Variable with respect to Minimum, Maximum, Mean, Standard Deviation of Values, Variance, and Coefficient of Variance

Variable	Minimum	Maximum	Mean	Std. Deviation	Variance	CV (%)
Age	24.000	65.000	47.460	11.156	124.458	23.506
Education	2.000	10.000	5.600	2.969	8.816	53.022
Functional education	3.000	9.000	4.560	2.032	4.129	44.561
Family size	2.000	16.000	4.780	2.306	5.318	48.244
Size of land holding (bigha)	0.700	22.000	5.789	4.615	21.300	79.723
No. of land fragments	1.000	9.000	5.820	2.173	4.722	37.337
Average size of land fragment (katha)	6.670	220.000	31.076	36.911	1362.446	118.778
Mean distance between two land fragments(m)	50.000	1000.000	353.160	240.345	57765.607	68.055
Size of homestead land (katha)	1.000	100.000	12.716	18.808	353.744	147.907
No. of crops cultivated	1.000	14.000	5.420	2.425	5.881	44.744
Total yield of crops cultivated (kg)	2354.000	106220.000	16862.280	22691.392	514899272.328	134.569
Total marketed surplus of crops cultivated (kg)	1640.000	103740.000	17745.985	20265.686	410698020.807	114.199
Total input cost (Rs)	17010.000	377050.000	100147.400	83623.776	6992935841.224	83.501
On farm income (Rs)	24000.000	1650000.000	179990.240	246451.347	60738266297.574	136.925
Family expenditure/year	36000.000	480000.000	138768.000	90016.083	8102895281.633	64.868
Savings/year	5000.000	350000.000	57700.000	59179.854	3502255102.041	102.565
Land under irrigation (bigha)	0.700	18.250	5.427	3.983	15.866	73.403
Cropping intensity (%)	60.000	441.670	158.059	79.747	6359.596	50.454

Variable	Minimum	Maximum	Mean	Std. Deviation	Variance	CV (%)
Total hours of irrigation given	188.000	3493.000	580.452	526.675	277386.685	90.735
Livestock count	0.000	21.000	2.840	3.862	14.913	135.975
Communication variable	75.000	250.000	125.260	42.223	1782.768	33.708
Consumption of coal, firewood, fuel wood (kg/year)	400.000	3840.000	1548.014	793.291	629310.010	51.246
Production of organic manure (kg/year)	54.750	25550.000	3967.019	3965.486	15725079.056	99.961

Result: The above table represents the distribution of 23 independent variables ($x1$-x_{23}) and 10 dependent variables (y_1-y_{10}) in terms of their range, mean, standard deviation and co-efficient of variance.

The distribution pattern of variable, **Number of land fragments (x_6)**, depicts that the maximum number of fragment is 9 and the minimum is 1. The mean number of fragment is 5.820 with standard deviation of 2.173. The coefficient of variance is 37.337 per cent which indicates the distribution of variable is fairly consistent.

The distribution pattern of variable, **Average size of land fragment (x_7)**, depicts that the maximum size of fragment is 220 katha and the minimum is 6.670 katha. The mean size of fragment is 31.076 katha with standard deviation of 36.911 katha. The coefficient of variance is 118.778 per cent which indicates the distribution of variable is less consistent.

The distribution pattern of variable, **Mean distance between two land fragments (x_8)**, depicts that the maximum distance between two fragments is 1000 m and theminimum is 50 m. The mean distance between two fragments is 353.160 mwith standard deviation of 240.345 m. The coefficient of variance is 68.055 per cent which indicates the distribution of variable is fairly consistent.

The distribution pattern of variable, **Size of homestead land (x_9)**, depicts that the maximum size is 100 katha and the minimum is 1 katha. The mean size of homestead land is 12.716katha with standard deviation of 18.808 katha. The coefficient of variance is 147.907 per cent which indicates the distribution of variable is more or less inconsistent.

The distribution pattern of variable, **No. of crops cultivated (x_{10})**, depicts that the maximum no. of crops cultivated throughout the year is 14 and the minimum is 1. The mean of no. of crops cultivated is 5.420 with standard deviation of 2.425. The coefficient of variance is 44.744 per cent which indicates the distribution of variable is fairly consistent.

The distribution pattern of variable, **Total yield of crops cultivated (x_{11})**, depicts that the maximum total yield of crops cultivated is 106220 kg and the minimum is 2354 kg. The mean of total yield of crops cultivated is 16862.280 kgwith standard deviation of 22691.392 kg. The coefficient of variance is 134.569 per cent which indicates the distribution of variable is less consistent.

The distribution pattern of variable, **Total marketed surplus of crops cultivated (x_{12})**, depicts that the maximum total marketed surplus of crops cultivated is 103740 kg and the minimum is 1640 kg. The mean of total marketed surplus of crops cultivated is 17745.985 kg with standard deviation of 20265.686 kg. The coefficient of variance is 114.199 per cent which indicates the distribution of variable is less consistent.

The distribution pattern of variable, **Total input cost (x_{13})**, depicts that the maximum total input cost is 377050 rupees and the minimum is 17010 rupees. The mean of total input cost is 100147.400 rupees with standard deviation of 83623.776 rupees. The coefficient of variance is 83.501 per cent which indicates the distribution of variable is fairly consistent.

The distribution pattern of variable, **on farm income (x_{14})**, depicts that the maximum on farm income is 1650000 rupees and the minimum is 24000 rupees. The mean on farm income is 179990.240 rupees with standard deviation of 246451.347 rupees. The coefficient of variance is 136.925 per cent which indicates the distribution of variable is less consistent.

The distribution pattern of variable, **Family expenditure/year (x_{15})**, depicts that the maximum family expenditure/year is 480000 rupees and the minimum is 36000 rupees. The mean family expenditure/year is 138768 rupees with standard deviation of 90016.08 rupees. The coefficient of variance is 64.868 per cent which indicates the distribution of variable is fairly consistent.

The distribution pattern of variable, **Savings/year (x_{16})**, depicts that the maximum savings/year is 350000 rupees and the minimum is 5000 rupees. The mean savings/year is 57700 rupees with standard deviation of 59179.85 rupees. The coefficient of variance is 102.565 per cent which indicates the distribution of variable is less consistent.

The distribution pattern of variable, **Land under irrigation (x_{17})**, depicts that the maximum land under irrigation is 18.250 bigha and the minimum is 0.700 bigha. The mean land under irrigation is 5.427 bigha with standard deviation of 3.983 bigha. The coefficient of variance is 73.403 per cent which indicates the distribution of variable is fairly consistent.

The distribution pattern of variable, **Cropping intensity (x_{18})**, depicts that the maximum crop intensity is 441.670 per cent and the minimum is 60 per cent. The mean crop intensity is 158.059 per cent with standard deviation of 79.747 per cent. The coefficient of variance is 50.454 per cent which indicates the distribution of variable is fairly consistent.

The distribution pattern of variable, **Total hours of irrigation given (x_{19})**, depicts that the maximum total hours of irrigation given is 3493 hours and the minimum is 188 hours. The mean of total hours of irrigation given is 580.452 hours with the standard deviation of 526.675hours. The coefficient of variance is 90.735 per cent which indicates the distribution of variable is fairly consistent.

The distribution pattern of variable, **Livestock count (x_{20})**, depicts that the maximum livestock count is 21 and the minimum is 0. The mean of livestock count is 2.840 with the standard deviation of 3.862. The coefficient of variance is 135.975 per cent which indicates the distribution of variable is less consistent.

The distribution pattern of variable, **Communication variable (x_{21})**, depicts that the maximum communication variable 250 and the minimum 75. The mean of communication variable is 125.260 with the standard deviation of 42.223. The coefficient of variance is 33.708 per cent which indicates the distribution of variable is fairly consistent.

The distribution pattern of variable, **Consumption of coal, firewood, fuel wood (x_{22})**, depicts that the maximum consumption of coal, firewood, fuel wood in a year is 3840 kg and the minimum 400 kg. The mean of consumption of coal, firewood, fuel wood is 1548.014 kg with the standard deviation of 793.291 kg. The coefficient of variance is 51.246 per cent which indicates the distribution of variable is fairly consistent.

The distribution pattern of variable, **Production of organic manure (x_{23})**, depicts that the maximum production of organic manure in a year is 25550 kg and the minimum 54.750 kg. The mean production of organic manure is 3967.019 kg with the standard deviation of 3965.486 kg. The coefficient of variance is 99.961 per cent which indicates the distribution of variable is fairly consistent.

Table-5.2: Coefficient of Correlation: Application of organic manure (y_1) and 23 independent variables

Sl. No.	Independent Variables	'r' Value	Remarks
1	Age (x_1)	0.071	
2	Education (x_2)	0.130	
3	Functional education (x_3)	-0.124	
4	Family size (x_4)	0.113	
5	Size of land holding (bigha) (x_5)	-0.244	
6	No. of land fragments (x_6)	**-0.446**	**
7	Average size of land fragment (katha) (x_7)	-0.120	
8	Mean distance between two land fragments (m) (x_8)	-0.025	
9	Size of homestead land (katha) (x_9)	0.105	
10	No. of crops cultivated (x_{10})	-0.150	
11	Total yield of crops cultivated (kg) (x_{11})	-0.116	
12	Total marketed surplus of crops cultivated (kg) (x_{12})	-0.078	
13	Total input cost (Rs) (x_{13})	0.010	
14	On farm income (Rs) (x_{14})	**0.438**	**
15	Family expenditure/year (x_{15})	0.138	
16	Savings/year (x_{16})	0.059	
17	Land under irrigation (bigha) (x_{17})	-0.226	
18	Cropping intensity (%) (x_{18})	-0.064	
19	Total hours of irrigation given (x_{19})	-0.074	
20	Livestock count (x_{20})	**0.916**	**

Sl. No.	Independent Variables	'r' Value	Remarks
21	Communication variable (x_{21})	0.023	
22	Consumption of coal,firewood,fuelwood (kg/year) (x_{22})	0.136	
23	Production of organic manure (kg/year) (x_{23})	**0.330**	*

**Correlation is significant at the 0.01 level

*Correlation is significant at the 0.05 level

Model-6.1: Coefficient of correlation: Application of organic manure (y_1) and independent variables having significant correlation.

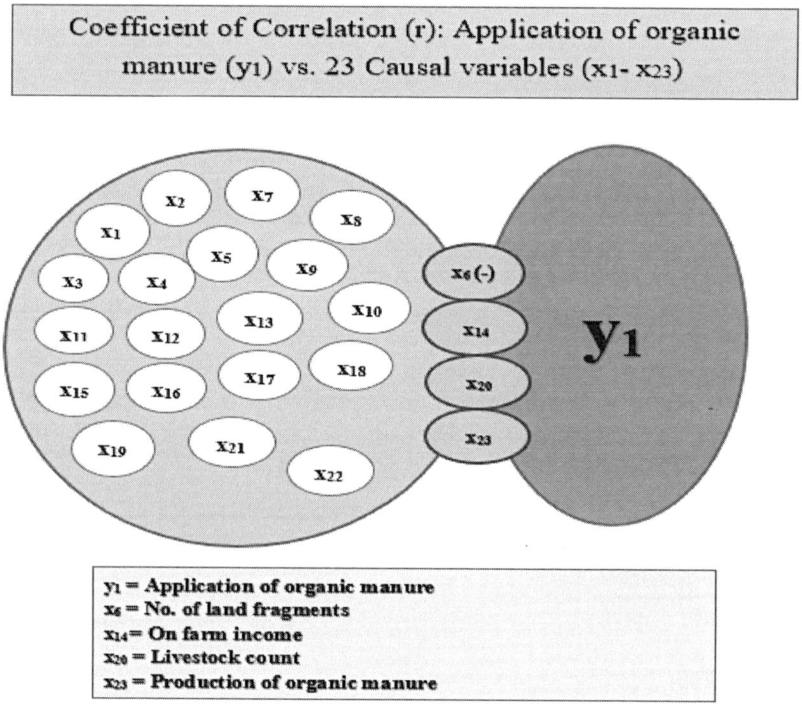

Result: Table-5.2 presents the coefficient of correlation between application of organic manure (y_1) and 23 independent variables. It has been found that the following variables on farm income (Rs) (x_{14}), livestock count (x_{20}) and production of organic manure (kg/year) (x_{23}) have been recorded significant and positive correlation with the dependent variable, while no. of land fragments (x_6) has recorded significant but negative correlation with the dependent variable, application of organic manure.

Revelation: The variable no. of land fragments (x_6) has recorded significant but negative correlation with application of organic manure (y_1). It is well discernible that when no. of land fragment is increasing, the application of organic manure has been reduced. The more is the no. of land fragments, the complex is the land

and resource management. If the no. of land fragments is more, then another possibility is that some few fragments are taken care of for better management while others are left aside.

The variable on farm income (x_{14}) has recorded positive but significant correlation with application of organic manure (y_1). The relation is obvious because whenever farm income is generating more, the intensity of management and caring goes up. The respondents, who are experiencing more income from their farms, they are becoming more sensitive to the soil health and land care and application of organic manures has been up scaled.

The variable livestock count (x_{20}) has recorded a significant and positive correlation with application of organic manure (y_1). This relationship is quite obvious because the primary source for organic manures for the farmers in this region are the count of livestock and crop residues. Whenever the count of livestock is increasing, the application of organic manure in their own lands is also increasing simultaneously.

The variable production of organic manure (x_{23}) has recorded a strong correlation with application of organic manure (y_1). The relationship is quite discernible simply because whenever at house and at farms production of organic manure is increasing the application of organic manure will also be increasing proportionately.

Table 5.3: Multiple Regression Analysis: Application of organic manure (y_1) Vs. 23 Causal Variables

Sl. No.	Variables	Reg. Coef. B	S.E. B	Beta	t Value
1	Age (x_1)	-3.406	9.605	-0.058	-0.355
2	Education (x_2)	98.716	51.968	0.283	1.900
3	Functional education (x_3)	-108.100	78.577	-0.299	-1.376
4	Family size (x_4)	-44.179	56.499	-0.103	-0.782
5	Size of land holding (bigha) (x_5)	-4.558	41.694	-0.014	-0.109
6	No. of land fragments (x_6)	-2.386	4.498	-0.108	-0.531
7	Average size of land fragment (katha) (x_7)	-0.280	0.631	-0.115	-0.444
8	Mean distance between two land fragments (m) (x_8)	21.762	14.515	0.295	1.499
9	Size of homestead land (katha) (x_9)	-132.406	76.095	-0.344	-1.740
10	No. of crops cultivated (x_{10})	0.008	0.014	0.115	0.572
11	Total yield of crops cultivated (kg) (x_{11})	0.009	0.007	0.192	1.280
12	Total marketed surplus of crops cultivated (kg) (x_{12})	0.003	0.001	0.284	2.399
13	Total input cost (Rs) (x_{13})	0.001	0.000	**0.382**	2.239

Sl. No.	Variables	Reg. Coef. B	S.E. B	Beta	t Value
14	On farm income (Rs) (x_{14})	0.003	0.002	0.239	1.646
15	Family expenditure/year (x_{15})	-0.001	0.002	-0.090	-0.700
16	Savings/year (x_{16})	-22.808	41.021	-0.136	-0.556
17	Land under irrigation (bigha) (x_{17})	2.842	1.399	**0.358**	2.031
18	Cropping intensity (%) (x_{18})	0.026	0.325	0.013	0.080
19	Total hours of irrigation given (x_{19})	180.378	43.116	**0.626**	4.184
20	Livestock count (x_{20})	-0.562	3.217	-0.030	-0.175
21	Communication variable (x_{21})	0.090	0.092	0.106	0.980
22	Consumption of coal,firewood,fuelwood (kg/year) (x_{22})	0.078	0.061	0.251	1.280
23	Production of organic manure (kg/ year) (x_{23})	-3.406	9.605	-0.058	-0.355

R square: 96.40%

The standard error of the estimate: 258.639

Result: Table-5.3 presents the full model of multiple regression analysis between 23 causal variables (x_1-x_{23}) and application of organic manure (y_1).

Revelation: The table-6.3 presents multiple regression analysis (full model) between 23 causal variables (x_1-x_{23}) and application of organic manure (y1). The three substantives impacts (Beta value) on consequent variable application of organic manure are in order.

Highlights of the result

- Beta coefficient of the causal variable total hours of irrigation given has been positive and significant. It implies that for proper decomposition of organic manure to transform it into available plant nutrient, a dedicated supply of water is must. A dedicated supply of water depends on total hours of irrigation given.
- Beta coefficient of the causal variable, total input cost, has been positive and significant. It implies that input cost has been reduced in a response to better application of organic manure. Whenever organic manure is applied to the soil, it will replace the need of application of chemical fertilizer, ecological resilience will be maintained and cost of input will be downsized and sustainability will be enhanced as well.
- Beta coefficient of the causal variable, land under irrigation, has also been positive and significant. It implies that in an irrigated agro-based ecosystem, the energy and input management system is very much sensitive and we have to be careful because number of irrigation involves energy. When

energy is more, the application of fertilizers will become more. At the same time when it is intensive agriculture under irrigated agro-based ecosystem, the total number of labour engaged will be more, means it has got energy higher equivalent. That is how it has got tremendous associating effect or ability to influence characters of other variable.

- The r^2 value stands at 96.40 per cent. It implies that combination of 23 causal variables has been quite justified, effective and able to explain 96.40 per cent of variance in the consequent variable application of organic manure.

Table 5.4: Stepwise Regression Analysis: Application of organic manure (y_1) Vs. 23 Causal Variables

Sl. No	Variables	Reg.coef. B	S.E. B	Beta	t value
1	Livestock count (x_{20})	165.152	22.540	0.574	7.327
2	Size of homestead land (katha) (x_9)	33.749	5.762	0.458	5.858
3	Size of land holding (bigha) (x_5)	-64.023	14.058	-0.381	-4.554
4	Production of organic manure (kg/year) (x_{23})	0.098	0.025	0.314	3.863
5	Education (x_2)	62.346	28.627	0.179	2.178

R square: 86.90%

The standard error of the estimate: 268.368

Model 5.2: Stepwise Regression Analysis: Application of organic manure (y_1)Vs. 23 Causal Variables

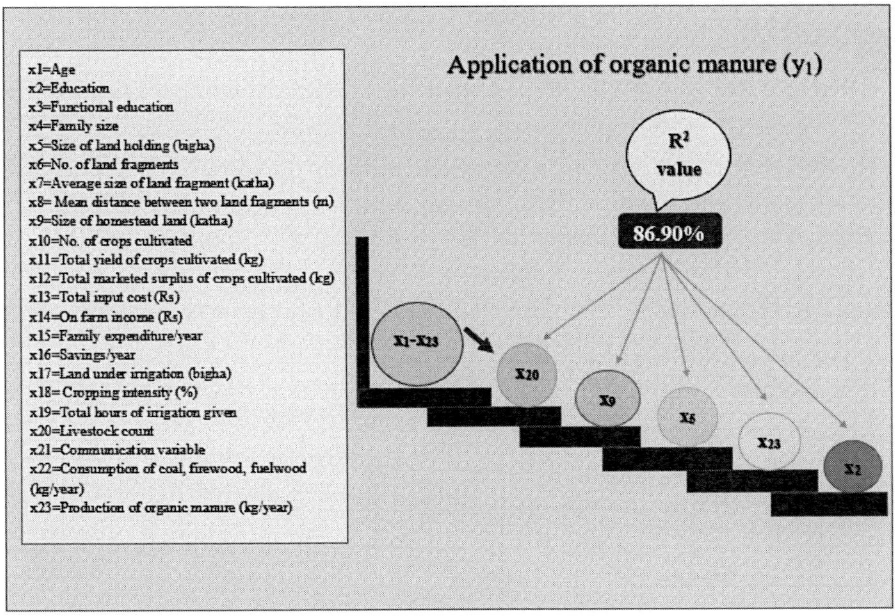

Result: Table 5.4 presents the stepwise regression analysis which elicits that five causal variables livestock count (x_{20}), size of homestead land (katha) (x_9), size of land holding (bigha) (x_5), production of organic manure (kg/year) (x_{23}) and education (x_2) has come out with stronger determining character on application of organic manure (y_1).

Revelation: It has been found that five causal variables have been retained at the last step of stepwise regression analysis. The variables are, livestock count, size of homestead land (katha), size of land holding (bigha), production of organic manure (kg/year) and education. These five causal variables together have contributed 86.90 per cent of variance. This indicates that the contribution of these causal variables has been substantive that rate of (23-5) i.e. 18 causal variables has contributed (96.40-86.90) per cent i.e. only 9.5 per cent. This indicates that to increase the application of organic manure, livestock count should be increased. Both the homestead land and size of land holding are conjointly contributing to the application of organic manure and soil health.

Homestead land plays a determining role for inviting and managing livestock resources and when it is managed at homestead ecology, the possibility of requirement of family members for livestock management grows high. Homestead land perhaps has not been able to draw much of attention from the agricultural scientists and ecologists. The social ecology of farming cannot be possible without comprehensive analysis and interpretation of homestead land ecology and its dynamics. The obvious advantages owning a big and organized homestead land is that it provokes and promotes horticulture, kitchen garden, livestock, fishery and poultry enterprises. Perhaps most or all of these enterprises generate a steady income as well as dedicated volume of organic residues that can be applied in other components of the farming.

Education has also come out as one of the crucial determinants in scaling up organic characters of farming.

These five variables have got a strong and disciple impact from the consequence of organic manure in the operative form for both the restoration of soil health and ecological management of farming. This five variables vis-a-vis should be dealt with attention and management skills.

Table 5.5: Path Analysis: Decomposition of Total Effect into Direct, Indirect and Residual Effect: Application of organic manure (y_1)

Sl. No	Variables	Total Effect	Direct Effect	Indirect Effect	Highest Indirect Effect
1	Age(x_1)	0.071	-0.058	0.129	0.048(x_5)
2	Education(x_2)	0.130	0.282	-0.152	-0.169(x_5)
3	Functional education(x_3)	-0.124	-0.299	0.175	0.060(x_5)
4	Family size(x_4)	0.113	-0.103	0.216	-0.130(x_5)
5	Size of land holding (bigha)(x_5)	-0.244	-0.014	-0.230	0.831(x_{17})

Sl. No	Variables	Total Effect	Direct Effect	Indirect Effect	Highest Indirect Effect
6	No. of land fragments(x_6)	-0.446	-0.108	-0.338	-0.742(x_5)
7	Average size of land fragment (katha)(x_7)	-0.120	-0.115	-0.005	-0.754(x_5)
8	Mean distance between two land fragments (m)(x_8)	-0.025	0.293	-0.318	-0.311(x_5)
9	Size of homestead land (katha) (x_9)	0.105	-0.344	0.449	-0.138(x_5)
10	No. of crops cultivated	-0.150	0.114	-0.264	0.194(x_{18})
11	Total yield of crops cultivated (kg)	-0.116	0.192	-0.308	0.701(x_5)
12	Total marketed surplus of crops cultivated (kg)	-0.078	0.284	-0.362	0.790(x_5)
13	Total input cost (Rs)	0.010	0.382	-0.372	0.134(x_9)
14	On farm income (Rs)	0.438	0.237	0.201	-0.174(x_5)
15	Family expenditure/year	0.138	-0.090	0.228	-0.383(x_5)
16	Savings/year	0.059	-0.136	0.195	-0.832(x_5)
17	Land under irrigation (bigha)	-0.226	0.358	-0.584	-0.988(x_5)
18	Cropping intensity (%)	-0.064	0.013	-0.077	0.170(x_5)
19	Total hours of irrigation given	-0.074	**0.626**	-0.700	-0.958(x_5)
20	Livestock count	0.916	-0.030	**0.946**	0.425(x_5)
21	Communication variable	0.023	0.106	-0.083	-0.180(x_5)
22	Consumption of coal, firewood, fuelwood (kg/year)	0.136	0.251	-0.115	0.491(x_5)
23	Production of organic manure (kg/year)	0.330	-0.055	0.385	0.239(x_5)

Residual effect: 0.036

Result: Table 5.5 presents the path analysis wherein the total effect of exogenous variable on consequent variable has been decomposed into direct, indirect and residual effect.

Revelation: It has been found that the variable total hour of irrigation given has exerted the highest direct effect. This implies that for proper decomposition of organic manure to transform it into available plant nutrient; a dedicated supply of water is must. A dedicated supply of water depends on total hours of irrigation given.

The count of livestock is the important determinant for the volume of organic manure to be produced by the farmer. The real concern in West Bengal agriculture is that the count of livestock is alarmingly decline hence natural source of organic manure in the form of cow dung has been jeopardized i.e. how in order to escalate the possibility and performance of CA, the count of livestock must have to enhanced otherwise that the restoration and augmentation of ecological resilience will remain illusion.

The size of the farm indicates the land resources and it also implies the resource endowment of the farmer. The higher is the size of holding, the bigger would be the investment ability and risk absorbing capability and that is how this variable has figure up as many as 20 out of 23 times to correct ionized the consequent variable application of organic manure.

The residual value is 0.036 it infers that a little over 3 per cent cannot be explaining with this combination of 23 variables. It has been supported by the R square value 96.40 per cent as well.

Table 5.6: Coefficient of Correlation: Rice stubble height left in the field (y_2) and 23 independent variables

Sl. No.	Independent Variables	'r' Value	Remarks
1	Age (x_1)	-0.280	*
2	Education(x_2)	-0.197	
3	Functional education(x_3)	-0.238	
4	Family size(x_4)	0.075	
5	Size of land holding (bigha) (x_5)	-0.092	
6	No. of land fragments (x_6)	-0.149	
7	Average size of land fragment (katha) (x_7)	-0.101	
8	Mean distance between two land fragments (m) (x_8)	0.182	
9	Size of homestead land (katha) (x_9)	0.100	
10	No. of crops cultivated (x_{10})	-0.144	
11	Total yield of crops cultivated (kg) (x_{11})	-0.151	
12	Total marketed surplus of crops cultivated (kg) (x_{12})	0.285	*
13	Total input cost (Rs) (x_{13})	0.604	**
14	On farm income (Rs) (x_{14})	-0.086	
15	Family expenditure/year (x_{15})	-0.164	
16	Savings/year (x_{16})	0.027	
17	Land under irrigation (bigha) (x_{17})	0.042	
18	Cropping intensity (%) (x_{18})	-0.255	
19	Total hours of irrigation given (x_{19})	-0.116	
20	Livestock count (x_{20})	-0.010	
21	Communication variable (x_{21})	-0.026	
22	Consumption of coal, firewood, fuelwood (kg/year) (x_{22})	0.016	
23	Production of organic manure (kg/year) (x_{23})	0.176	

**Correlation is significant at the 0.01 level , *Correlation is significant at the 0.05 level

Model 5.3: Coefficient of correlation: Rice stubble height left in the field (y_2) and independent variables having significant correlation.

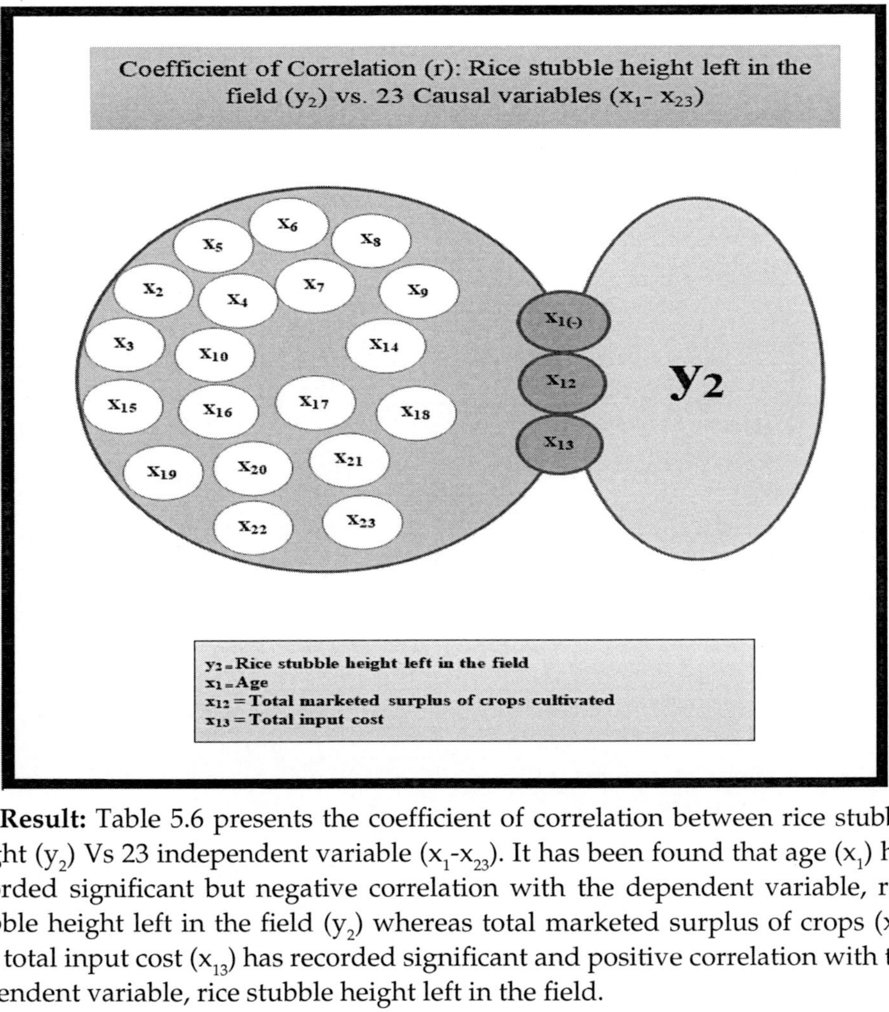

Result: Table 5.6 presents the coefficient of correlation between rice stubble height (y_2) Vs 23 independent variable (x_1-x_{23}). It has been found that age (x_1) has recorded significant but negative correlation with the dependent variable, rice stubble height left in the field (y_2) whereas total marketed surplus of crops (x_{12}) and total input cost (x_{13}) has recorded significant and positive correlation with the dependent variable, rice stubble height left in the field.

Revelation: The results justify that for the young farmers, the left out stubble heights are on higher sides. So, the concept of ecological resilience is more reflected in the younger farmers. The younger respondents, those who have better education and higher school going experiences, they are increasingly exposed to ecological training and education. They are somehow attending to ecological concerns of their farming.

It is interesting to note that the total marketed surplus of crops cultivated has shown a positive change when the higher volume of crop residues left in the field. So, the recycling of crop residue has started yielding positive impact on the economy of farming through of generation of marketed surplus.

Total input cost does not show any positive response for rationalizing cost of cultivation by registering a negative correlation with total volume of crop residue left in the field. Has it been a negative correlation between these two variables, we could infer that residue management has got economically beneficial and has got decisive impact on its dividend.

Table 5.7: Multiple Regression Analysis: Rice stubble height left in the field (y_2) Vs. 23 Causal Variables

Sl. No.	Variables	Reg. Coef. B	S.E. B	Beta	t Value
1	Age (x_1)	-0.033	0.032	-0.147	-1.014
2	Education (x_2)	0.300	0.174	0.228	1.721
3	Functional education (x_3)	-0.298	0.263	-0.219	-1.131
4	Family size (x_4)	-0.105	0.189	-0.065	-0.554
5	Size of land holding (bigha) (x_5)	-0.049	0.140	-0.040	-0.349
6	No. of land fragments (x_6)	0.005	0.015	0.058	0.319
7	Average size of land fragment (katha) (x_7)	0.003	0.002	**0.313**	1.350
8	Mean distance between two land fragments (m) (x_8)	-0.076	0.049	-0.275	-1.564
9	Size of homestead land (katha) (x_9)	0.269	0.255	0.186	1.055
10	No. of crops cultivated (x_{10})	2.867E-05	0.000	0.108	0.607
11	Total yield of crops cultivated (kg) (x_{11})	3.517E-05	0.000	0.203	1.512
12	Total marketed surplus of crops cultivated (kg) (x_{12})	3.831E-05	0.000	**1.059**	10.026
13	Total input cost (Rs) (x_{13})	-3.147E-07	0.000	-0.036	-0.239
14	On farm income (Rs) (x_{14})	1.722E-06	0.000	0.031	0.243
15	Family expenditure/year (x_{15})	-5.707E-06	0.000	-0.105	-0.912
16	Savings/year (x_{16})	-0.016	0.138	-0.026	-0.118
17	Land under irrigation (bigha) (x_{17})	-0.010	0.005	**-0.325**	-2.064
18	Cropping intensity (%) (x_{18})	-0.001	0.001	-0.114	-0.813
19	Total hours of irrigation given (x_{19})	0.106	0.145	0.097	0.730
20	Livestock count (x_{20})	-0.011	0.011	-0.161	-1.052
21	Communication variable (x_{21})	0.001	0.000	0.252	2.599
22	Consumption of coal, firewood, fuel wood (kg/year) (x_{22})	7.593E-05	0.000	0.065	0.372
23	Production of organic manure (kg/year) (x_{23})	-0.033	0.032	-0.147	-1.014

R square: 97.20%

The standard error of the estimate: 0.867

Result: Table-5.7 presents the full model of multiple regression analysis between 23 causal variables (x_1-x_{23}) and rice stubble height left in the field (y_2).

Revelation: The table-6.7 presents multiple regression analysis (full model) between 23 causal variables (x_1-x_{23}) and rice stubble height left in the field (y_2). The three substantives impact (Beta value) on consequent variable rice stubble height left in the field are in order.

Highlights of the result

- Beta coefficient of the causal variable total marketed surplus of crops cultivated, has been positive and significant. It implies that whenever the total marketed surplus of crops is higher, it is expected to take care of the land by leaving higher volume of stubble in the field.
- Beta coefficient of the causal variable, land under irrigation, has been negative but significant. It implies that in an irrigated agro-based ecosystem the energy and input management system is very much sensitive and we have to be careful because number of irrigation involves energy. When energy is more, then the application of fertilizers will become more. At the same time when it is intensive agriculture under irrigated agro-based ecosystem, the total number of labor engaged will be more, means it has got energy equivalent. So, that is how it has got tremendous associating effect or ability to influence characters of other variable.
- Beta coefficient of the causal variable, average size of land fragment, has been positive and significant. It implies that the average size of land fragment has come up as one of the strongest determinant to decide on the nature, volume and type of crop residue management. The average size of land fragment is immensely important in deciding about all kinds of farm management. The less is the average size of land fragments, the higher would be the energy loses that is how with the number of land fragments the whole of the farm operations will turn energy prodigal and cost intensive.

The r^2 value stands at 97.20 per cent. It implies that combination of 23 causal variables has been quite justified, effective and able to explained 97.20 per cent of variance in the consequent variable rice stubble height left in the field.

Table 5.8: Stepwise Regression Analysis: Rice stubble height left in the field (y_2) Vs. 23 Causal Variables

Sl. No	Variables	Reg.coef. B	S.E. B	Beta	t value
1	Total input cost (Rs)(x_{13})	3.085E-05	0.000	0.852	11.154
2	Age (x_1)	-0.076	0.017	-0.339	-4.438
3	Functional education (x_3)	-0.402	0.105	-0.296	-3.820

R square: 85.30%

The standard error of the estimate: 1.024

Model 5.4: Stepwise Regression Analysis: Rice stubble height left in the field (y_2) Vs. 23 Causal Variables

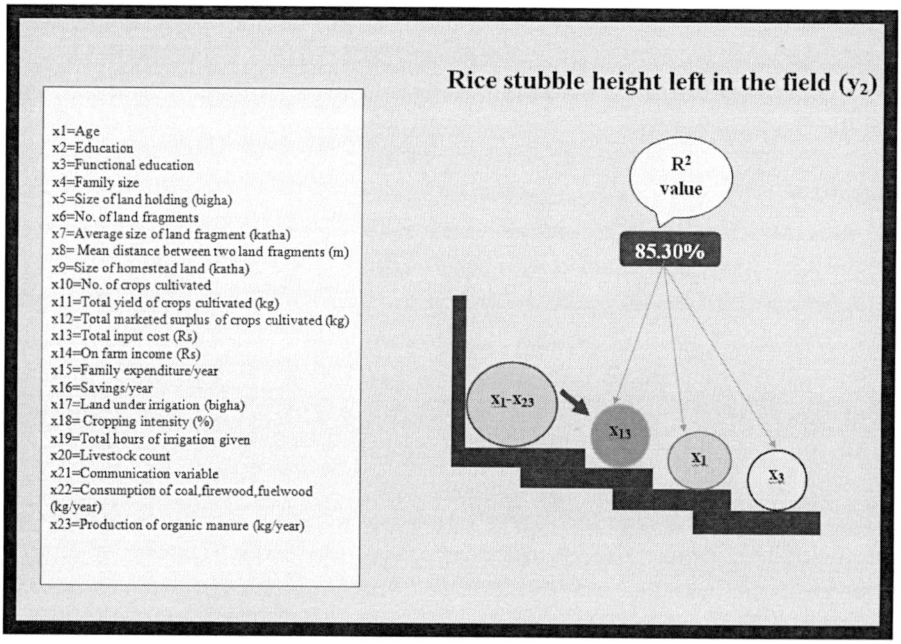

Result: Table 5.8 presents the stepwise regression analysis which elicits that three causal variables total input cost (x_{13}), age (x_1) and functional education (x_3) has come out with stronger determining character on the consequent variable, rice stubble height left in the field (y_2).

Revelation: Table 5.8 presents the stepwise regression analysis for screening out the three causal variables with the most significant functional contribution to the consequent variable, rice stubble height left in the field. It has been found that the following variable has been retained at the last step and these are total input cost, age and functional education. The r_2 value stands at 85.30 per cent which implies that these three variables collectively contributed 85.30 per cent of variance in the consequent variable rice stubble height left in the field. The rest twenty variables have contributed only (97.20-85.30) per cent i.e., 11.9 per cent variance.

The fact has elicited from this hard evidence speaks that amongst the young farmers the perception of conservation agriculture is gaining ground and it has been supported by other causal variable, functional education in the form of training, experiential learning or exposure visit in organic farming as well as conservation agriculture to ultimately socialize the concept of ecological agenda of agriculture.

Total input cost here has come out as an important determinant for characterizing the mode of application of crop residue in the field.

Table 5.9: Path Analysis: Decomposition of Total Effect into Direct, Indirect and Residual Effect: Rice stubble height left in the field (y_2)

Sl. No	Variables	Total Effect	Direct Effect	Indirect Effect	Highest Indirect Effect
1	Age (x_1)	-0.280	-0.147	-0.133	0.594(x_5)
2	Education (x_2)	-0.197	0.228	-0.425	-0.165(x_5)
3	Functional education (x_3)	-0.238	-0.219	-0.019	-0.781(x_5)
4	Family size (x_4)	0.075	-0.065	0.140	-0.122(x_5)
5	Size of land holding (bigha) (x_5)	-0.092	-0.040	-0.052	0.917(x_{17})
6	No. of land fragments (x_6)	-0.149	0.056	-0.205	-1.363(x_5)
7	Average size of land fragment (katha) (x_7)	-0.101	0.313	-0.414	-0.891(x_5)
8	Mean distance between two land fragments (m) (x_8)	0.182	-0.275	0.457	-0.078(x_5)
9	Size of homestead land (katha) (x_9)	0.100	0.186	-0.086	-0.088(x_5)
10	No. of crops cultivated (x_{10})	-0.144	0.108	-0.252	-0.170(x_{18})
11	Total yield of crops cultivated (kg) (x_{11})	-0.151	0.201	-0.352	0.286(x_5)
12	Total marketed surplus of crops cultivated (kg) (x_{12})	0.285	**1.059**	**-0.774**	0.451(x_5)
13	Total input cost (Rs) (x_{13})	0.604	-0.036	0.640	-0.199(x_9)
14	On farm income (Rs) (x_{14})	-0.086	0.031	-0.117	-0.989(x_5)
15	Family expenditure/year (x_{15})	-0.164	-0.105	-0.059	-0.539(x_5)
16	Savings/year (x_{16})	0.027	-0.026	0.053	-0.528(x_5)
17	Land under irrigation (bigha) (x_{17})	0.042	-0.325	0.367	-0.990(x_5)
18	Cropping intensity (%) (x_{18})	-0.255	-0.114	-0.141	0.616(x_5)
19	Total hours of irrigation given (x_{19})	-0.116	0.093	-0.209	-0.758(x_5)
20	Livestock count (x_{20})	-0.010	-0.161	0.151	0.780(x_5)
21	Communication variable (x_{21})	-0.026	0.252	-0.278	-0.330(x_5)
22	Consumption of coal,firewood,fuelwood (kg/year) (x_{22})	0.016	0.065	-0.049	0.037(x_5)
23	Production of organic manure (kg/year) (x_{23})	0.176	-0.146	0.324	0.275(x_5)

Residual effect: 0.033

Result: Table 5.9 presents the path analysis wherein the total effect of exogenous variable on consequent variable has been decomposed into direct, indirect and residual effect.

Revelation: Total marketed surplus of crops cultivated has exerted both the highest direct effect and highest indirect effect on rice stubble height left in the field. Whenever the total marketed surplus of crops is higher, it is expected to take care of the land by leaving higher volume of stubble in the field. At the same time, it has exerted highest indirect effect and it has been found that in cases the marketed surplus is less, then it is adversely affecting the prospects of stubble height or vis-a-vis the mixing of crop residue in the field.

The size of land holding has appeared as many as 20 out of 23 times in exerting the highest effect on rice stubble height left in the field whenever the size of farm is big. The farmers' precepts that it is advantageous to leave the more stubble on the field because it is difficult to manage the huge volume of straw and it also curb labor cost.

The residual value is 0.033 it infers that a little over 3 per cent cannot be explain with this combination of 23 variables. It has been supported by the R square value 97.20 per cent as well.

Table 5.10: Coefficient of Correlation: Crop residue left in the field (y_3) and 23 independent variables

Sl. No.	Independent Variables	'r' Value	Remarks
1	Age (x_1)	-0.310	*
2	Education (x_2)	0.004	
3	Functional education (x_3)	0.015	
4	Family size (x_4)	-0.089	
5	Size of land holding (bigha) (x_5)	0.157	
6	No. of land fragments (x_6)	0.162	
7	Average size of land fragment (katha) (x_7)	0.062	
8	Mean distance between two land fragments (m) (x_8)	0.218	
9	Size of homestead land (katha) (x_9)	0.078	
10	No. of crops cultivated (x_{10})	0.149	
11	Total yield of crops cultivated (kg) (x_{11})	0.097	
12	Total marketed surplus of crops cultivated (kg) (x_{12})	-0.018	
13	Total input cost (Rs) (x_{13})	0.007	
14	On farm income (Rs) (x_{14})	-0.071	
15	Family expenditure/year (x_{15})	0.156	
16	Savings/year (x_{16})	-0.078	
17	Land under irrigation (bigha) (x_{17})	0.200	
18	Cropping intensity (%) (x_{18})	-0.111	

Sl. No.	Independent Variables	'r' Value	Remarks
19	Total hours of irrigation given (x_{19})	0.074	
20	Livestock count (x_{20})	0.021	
21	Communication variable (x_{21})	0.617	**
22	Consumption of coal, firewood, fuelwood (kg/year) (x_{22})	-0.201	
23	Production of organic manure (kg/year) (x_{23})	-0.005	

**Correlation is significant at the 0.01 level

*Correlation is significant at the 0.05 level

Model 5.5: Coefficient of correlation: Crop residue left in the field (y_3) and independent variables having significant correlation.

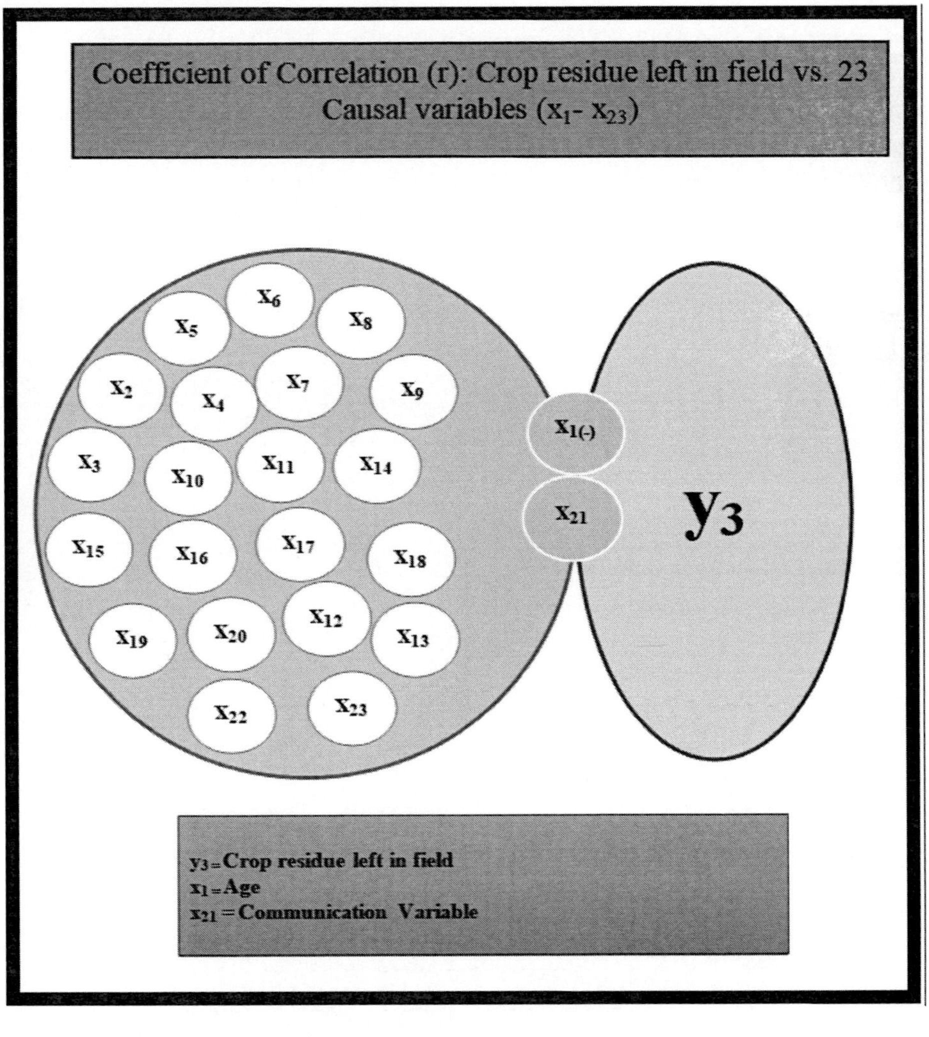

Result: Table 5.10 presents the coefficient of correlation between crop residue left in the field (y_3) Vs 23 independent variables (x_1-x_{23}). It has been found that the following variable age (x_1) has recorded significant but negative correlation and communication variable (x_{21}) has recorded significant and positive correlation with the dependent variable, crop residue left in the field (y_3).

Revelation: The young farmers are more prone to crop residue management than the older one through different sensitization and training programmes. The young farmers might have been exposed to ecological education and perception, attained proficiency in restoring the ecological resilience.

The role and contribution of different media, TV/radio, interpersonal opinion leader and cosmopolite source as fertilizer dealer, they have been sensitized about climate change or ecological resilience. Their responses have been recorded so that the yields of different rice variety over decade have been declined, this was their observation and perception. They have also observed and recorded during field survey, there has been a serious decline of biodiversity; decline of local cultivars is alarming and that is how and why through different educational programmes, exposure visits, training and sensitization programmes. They have built up a favourable attitudes and perception about crop residue management.

Table 5.11.1:Multiple Regression Analysis: Crop residue left in the field (y3) Vs. 23 Causal Variables

Sl. No.	Variables	Reg. Coef. B	S.E. B	Beta	t Value
1	Age (x1)	-32.999	17.038	-0.714	-1.937
2	Education (x2)	36.017	92.184	0.132	0.391
3	Functional education (x3)	-171.682	139.386	-0.609	-1.232
4	Family size (x4)	-130.143	100.222	-0.388	-1.299
5	Size of land holding (bigha) (x5)	52.108	73.959	0.207	0.705
6	No. of land fragments (x6)	-1.858	7.979	-0.108	-0.233
7	Average size of land fragment (katha) (x7)	-1.577	1.119	-0.831	-1.408
8	Mean distance between two land fragments (m) (x8)	43.918	25.748	0.763	1.706
9	Size of homestead land (katha) (x9)	54.540	134.983	0.182	0.404
10	No. of crops cultivated (x10)	-0.023	0.025	-0.412	-0.905
11	Total yield of crops cultivated (kg) (x11)	-0.008	0.012	-0.235	-0.686
12	Total marketed surplus of crops cultivated (kg) (x12)	-0.003	0.002	-0.380	-1.413
13	Total input cost (Rs) (x13)	-0.002	0.001	-0.851	-2.193

Sl. No.	Variables	Reg. Coef. B	S.E. B	Beta	t Value
14	On farm income (Rs) (x14)	0.001	0.004	0.132	0.399
15	Family expenditure/year (x15)	-0.004	0.003	-0.363	-1.244
16	Savings/year (x16)	137.116	72.766	1.044	1.884
17	Land under irrigation (bigha) (x17)	1.829	2.482	0.296	0.737
18	Cropping intensity (%) (x18)	-0.786	0.576	-0.487	-1.365
19	Total hours of irrigation given (x19)	106.871	76.482	0.475	1.397
20	Livestock count (x20)	10.337	5.706	0.704	1.812
21	Communication variable (x21)	0.090	0.163	0.136	0.552
22	Consumption of coal, firewood, fuelwood (kg/year) (x22)	-0.108	0.108	-0.446	-1.002
23	Production of organic manure (kg/year) (x23)	-32.999	17.038	-0.714	-1.937

R square: 81.60%

The standard error of the estimate: 458.792

Result: Table-5.11.1 presents the full model of multiple regression analysis between 23 causal variables (x_1-x_{23}) and crop residue left in the field (y_3).

Revelation: The table presents multiple regression analysis (full model) between 23 causal variables (x_1-x_{23}) and crop residue left in the field (y3). The three substantives impacts (Beta value) on consequent variable crop residue left in the field are in order.

Highlights of the result

- Beta coefficient of the causal variable total input cost has been negative but significant. It implies that input has been reduced in a response to better crop residue management. Whenever the crop residue is mix to the soil it will replace the need of application of chemical fertilizer then ecological resilience will be maintained and cost of input will be downsized and sustainability will be enhanced.
- The another highlights of this result is that with the reduction of input cost, savings have been encouraged which is quite obvious.
- The average size of land fragment has come up as one of the strongest determinant to decide on the nature, volume and type of crop residue management. The average size of land fragment is immensely important in deciding about all kinds of farm management. The less is the average size of land fragments, the higher would be the energy loses that is how with the number of land fragments the whole of the farm operations will turn energy prodigal and cost intensive.

- The r^2 value stands at 81.60 per cent. It implies that combination of 23 causal variables has explained 81.60 per cent of variance in the consequent variable crop residue left in the field.

Table 5.12: Stepwise Regression Analysis: Crop residue left in the field (y_3) Vs. 23 Causal Variables

Sl. No	Variables	Reg.coef. B	S.E. B	Beta	t value
1	Communication variable(x_{21})	7.455	2.231	0.508	3.341
2	Mean distance between two land fragments (m)(x_8)	0.629	0.288	0.332	2.183

R square: 37.60%

The standard error of the estimate: 430.337

Model 6.6: Stepwise Regression Analysis: Crop residue left in the field (y_3) Vs. 23 Causal Variable

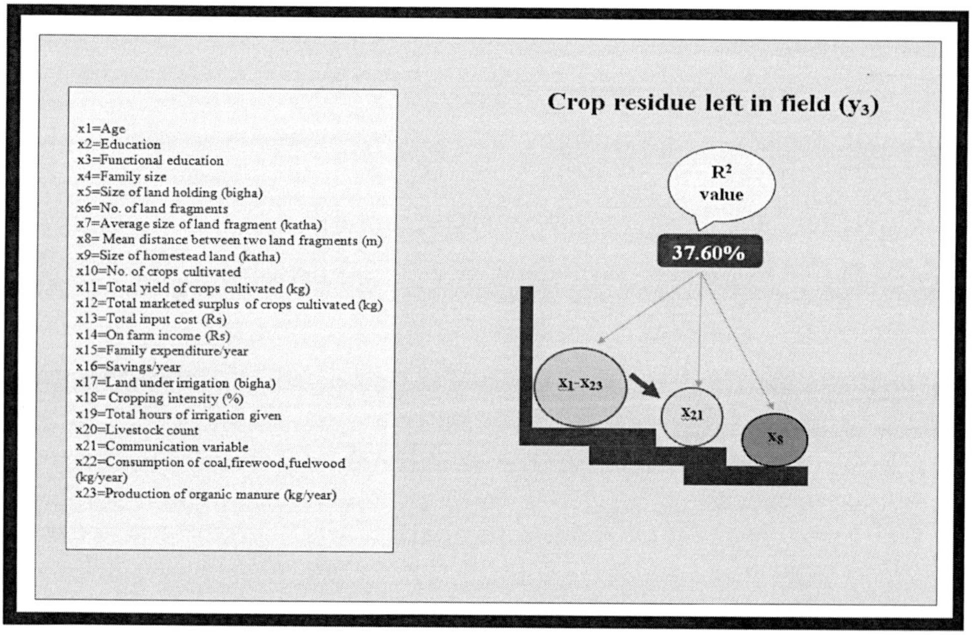

Result: Table 5.11 presents the stepwise regression analysis which elicits that two causal variables, communication variable(x_{21}) and mean distance between two land fragments (m)(x_8) has come out with stronger determining character on the consequent variable, crop residue left in the field (y_3).

Revelation: Table 5.11 presents the stepwise regression analysis for screening out the two causal variables with the most significant functional contribution to the consequent variable, crop residue left in the field. It has been found that the following variable has been retained at the last step and these are communication variable and mean distance between two land fragments. The r_2 value stands at 37.60 per cent which implies that these two variables collectively contributed 37.60 per cent of variance in the consequent variable crop residue left in the field. The rest twenty variables have contributed only (81.60-37.60) per cent i.e., 44 per cent variance.

The fact has elicited from this hard evidence speaks that the role and contribution of different media, TV/radio, interpersonal opinion leader and cosmopolite source as fertilizer dealer, they have been sensitized about climate change or ecological resilience. Their responses have been recorded so that the yields of different rice variety over decade have been declined, this was their observation and perception. They have also observed and recorded during field survey, there has been a serious decline of biodiversity; decline of local cultivars is alarming and that is how and why through different educational programmes, exposure visits, training and sensitization programmes. They have built up a favourable attitudes and perception about crop residue management.

Mean distance between two land fragments has come out as an important determinant for characterizing the crop residue management. It is expected that when distance will be reduced, the crop residue management would go better. On the other hand, it is also a fact that when land fragments are residing at higher distance most of the residues are not carried back in their home but it is mixed in the soil.

Table 5.13: Path Analysis: Decomposition of Total Effect into Direct, Indirect and Residual Effect: Crop residue left in the field (y_3)

Sl. No	Variables	Total Effect	Direct Effect	Indirect Effect	Highest Indirect Effect
1	Age (x_1)	-0.310	-0.714	0.404	-0.992(x_{17})
2	Education (x_2)	0.004	0.132	-0.128	0.397(x_{17})
3	Functional education (x_3)	0.015	-0.609	0.624	0.124(x_{17})
4	Family size (x_4)	-0.089	-0.388	0.299	0.507(x_{17})
5	Size of land holding (bigha) (x_5)	0.157	0.207	-0.050	0.348(x_{17})
6	No. of land fragments (x_6)	0.162	-0.108	0.270	0.383(x_{17})
7	Average size of land fragment (katha) (x_7)	0.062	-0.831	**0.893**	0.163(x_{17})
8	Mean distance between two land fragments (m) (x_8)	0.218	0.761	-0.543	0.512(x_{17})
9	Size of homestead land (katha) (x_9)	0.078	0.182	-0.104	0.716(x_{17})
10	No. of crops cultivated (x_{10})	0.149	-0.410	0.559	0.424(x_{17})
11	Total yield of crops cultivated (kg) (x_{11})	0.097	-0.235	0.332	-0.136(x_{17})
12	Total marketed surplus of crops cultivated (kg) (x_{12})	-0.018	-0.380	0.362	-0.666(x_{17})

Sl. No	Variables	Total Effect	Direct Effect	Indirect Effect	Highest Indirect Effect
13	Total input cost (Rs) (x_{13})	0.007	-0.851	0.858	-0.389(x_{17})
14	On farm income (Rs) (x_{14})	-0.071	0.132	-0.203	0.831(x_{17})
15	Family expenditure/year (x_{15})	0.156	-0.361	0.517	0.130(x_{17})
16	Savings/year (x_{16})	-0.078	**1.044**	-1.122	0.913(x_{17})
17	Land under irrigation (bigha) (x_{17})	0.200	0.296	-0.096	-0.428(x_5)
18	Cropping intensity (%) (x_{18})	-0.111	-0.487	0.376	-0.665(x_{17})
19	Total hours of irrigation given (x_{19})	0.074	0.475	-0.401	0.585(x_{17})
20	Livestock count (x_{20})	0.021	0.702	-0.681	-0.474(x_{17})
21	Communication variable (x_{21})	0.617	0.136	0.481	-1.033(x_5)
22	Consumption of coal,firewood,fuelwood (kg/year) (x_{22})	-0.201	-0.446	0.245	-0.837(x_{17})
23	Production of organic manure (kg/ year) (x_{23})	-0.005	-0.713	0.708	-0.282(x_{17})

Residual effect: 0.176

Result: Table 5.13 presents the path analysis wherein the total effect of exogenous variable on consequent variable has been decomposed into direct, indirect and residual effect.

Revelation: When the crop residue management is done properly, then the soil fertility status becomes better which ensures better yield as well as the economic status of the farmers. Therefore, the direct effect of savings has been found to exert highest positive effect on crop residue left in the field.

The average size of land fragment is immensely important in all kinds of farm management. It becomes labour intensive, cost and energy prodigal that is why this variable has shown highest indirect effect on crop residue left in the field.

The residual value is 0.176 it infers that a little over 17.6 per cent cannot be explain with this combination of 23 variables. It has been supported by the R square value 81.60 per cent as well.

It has been found that the exogenous variable land under irrigation has appeared as many as 21 out of 23 times in exerting the highest effect on the consequent variable crop residue left in the field, it implies that in an irrigated agro-based ecosystem the crop residue left in the field are higher. It is obvious that in a moist soil condition, sometimes it is muddy and therefore the general tendency is to leave higher proportion of crop residue left in the field.

Two reasons are –i) In a wet condition, it is difficult to collect, ii) In a wet muddy condition, it is easy to admix with the soil which ultimately would add to the fertility of the soil.

Table 5.14: Coefficient of Correlation (r): Climate resilient technology (y_4) and 23 independent variables

Sl. No.	Independent Variables	'r' Value	Remarks
1	Age (x_1)	-0.137	
2	Education (x_2)	-0.181	
3	Functional education (x_3)	0.153	
4	Family size (x_4)	-0.171	
5	Size of land holding (bigha) (x_5)	-0.083	
6	No. of land fragments (x_6)	0.167	
7	Average size of land fragment (katha) (x_7)	-0.253	
8	Mean distance between two land fragments (m) (x_8)	-0.246	
9	Size of homestead land (katha) (x_9)	-0.189	
10	No. of crops cultivated (x_{10})	0.101	
11	Total yield of crops cultivated (kg) (x_{11})	0.074	
12	Total marketed surplus of crops cultivated (kg) (x_{12})	0.090	
13	Total input cost (Rs) (x_{13})	0.027	
14	On farm income (Rs) (x_{14})	-0.019	
15	Family expenditure/year (x_{15})	-0.205	
16	Savings/year (x_{16})	0.148	
17	Land under irrigation (bigha) (x_{17})	0.031	
18	Cropping intensity (%) (x_{18})	-0.064	
19	Total hours of irrigation given (x_{19})	-0.151	
20	Livestock count (x_{20})	-0.250	
21	Communication variable (x_{21})	0.004	
22	Consumption of coal,firewood,fuelwood (kg/year) (x_{22})	**0.387**	*
23	Production of organic manure (kg/year) (x_{23})	**-0.359**	*

**Correlation is significant at the 0.01 level

*Correlation is significant at the 0.05 level

Model 5.7: Coefficient of correlation: Climate resilient technology (y_4) and independent variables having significant correlation.

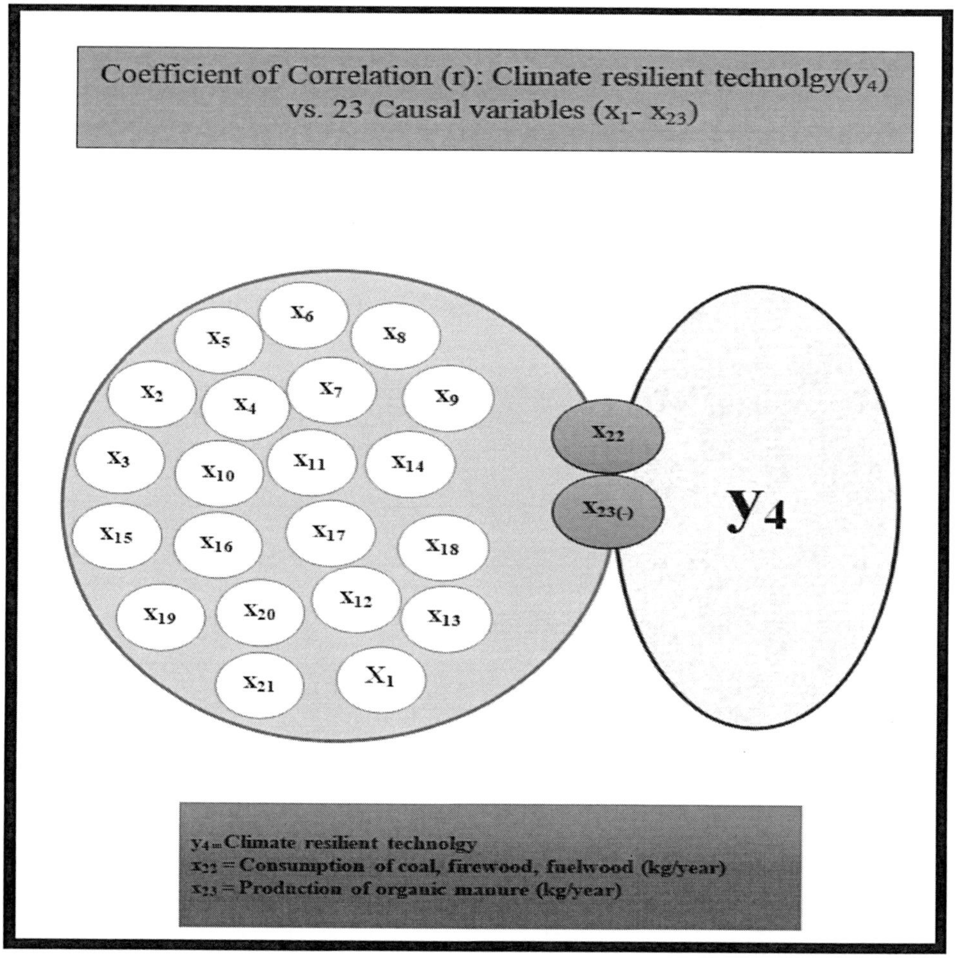

Result: Table 5.14 presents the coefficient of correlation between climate resilient technology (y_4) Vs 23 independent variables (x_1-x_{23}). It has been found that the following variable consumption of coal, firewood, fuel wood (kg/year) (x_{22}) has recorded significant but positive correlation and production of organic manure (kg/year) (x_{23}) has recorded significant and negative correlation with the dependent variable, climate resilient technology (y_4).

Revelation: In an attempt to make our farming ecological resilient and productive, we have to be sensitive and careful about energy management which again is missing badly in our farming. We are mostly concerned about application of fertilizer, drenching of irrigation water, indiscriminate application of pesticides and sometimes unaudited mechanization as well. We mostly remain callous and careless about the energy aspect of the farm. Our farms are biologically productive and energy wise prodigal that is why the variable consumption of coal, firewood, fuelwood, which in turn indicates the energy behavior which has got tremendous effect on climate resilient agriculture. If we want to make our farm climate resilient, eco-friendly and ecologically productive, then we have to rationalize the consumption of energy. Sometimes whole of a night, the farm remains under operating pump-set and the farmer remains careless about the energy loses contributing climate change and global warming.

Production of organic manure has been so important in replacing chemical fertilizer, regaining ecological resilience and re-engineering the production process perfectly tuned to ecological functions. When production process goes in compliance with ecological principle and prerequisite, the whole of the production system is certain to go climate resilient.

Table 5.15: Multiple Regression Analysis: Climate resilient technology (y_4) Vs. 23 Causal Variables

Sl. No.	Variables	Reg. Coef. B	S.E. B	Beta	t Value
1	Age (x_1)	-0.043	0.026	-0.573	-1.666
2	Education (x_2)	0.288	0.140	0.647	2.050
3	Functional education (x_3)	-0.459	0.212	**-0.996**	-2.162
4	Family size (x_4)	-0.145	0.153	-0.265	-0.950
5	Size of land holding (bigha) (x_5)	-0.123	0.113	-0.300	-1.095
6	No. of land fragments (x_6)	-0.003	0.012	-0.092	-0.214
7	Average size of land fragment (katha) (x_7)	-0.004	0.002	**-1.277**	-2.319
8	Mean distance between two land fragments (m) (x_8)	0.057	0.039	0.607	1.456
9	Size of homestead land (katha) (x_9)	-0.136	0.206	-0.277	-0.661
10	No. of crops cultivated (x_{10})	-4.475E-05	0.000	-0.499	-1.176
11	Total yield of crops cultivated (kg) (x_{11})	1.159E-06	0.000	0.020	0.062
12	Total marketed surplus of crops cultivated (kg) (x_{12})	3.238E-06	0.000	0.264	1.052
13	Total input cost (Rs) (x_{13})	-9.666E-07	0.000	-0.330	-0.912
14	On farm income (Rs) (x_{14})	7.064E-07	0.000	0.038	0.124
15	Family expenditure/year (x_{15})	-6.755E-06	0.000	-0.365	-1.340
16	Savings/year (x_{16})	0.166	0.111	0.772	1.494
17	Land under irrigation (bigha) (x_{17})	0.003	0.004	0.258	0.689
18	Cropping intensity (%) (x_{18})	-0.001	0.001	-0.462	-1.387
19	Total hours of irrigation given (x_{19})	0.104	0.116	0.283	0.892

Sl. No.	Variables	Reg. Coef. B	S.E. B	Beta	t Value
20	Livestock count (x_{20})	0.005	0.009	0.204	0.563
21	Communication variable (x_{21})	0.001	0.000	0.649	2.823
22	Consumption of coal,firewood,fuelwood (kg/year) (x_{22})	-0.001	0.000	**-1.270**	-3.059
23	Production of organic manure (kg/year) (x_{23})	-0.043	0.026	-0.573	-1.666

R square: 84.00%

The standard error of the estimate: 0.699

Result: Table-5.15 presents the full model of multiple regression analysis between 23 causal variables (x_1-x_{23}) and climate resilient technology (y_4).

Revelation: The table presents multiple regression analysis (full model) between 23 causal variables (x_1-x_{23}) and climate resilient technology (y_4). The three substantives impacts (Beta value) on consequent variable climate resilient technology are in order.

Highlights of the result

- Beta coefficient of the causal variable average size of land fragment is negative but significant, it implies that the average size of land fragment has contributed substantially on climate resilient technology. The more is the number of land fragment, the higher would be the energy loses that is how with the number of fragments the whole of the farm operations will turn energy prodigal and cost intensive.
- Beta coefficient of the causal variable consumption of coal, firewood, fuelwood is negative but significant, it implies that the energy consumption behavior and pattern of the respondent has got substantive effect on the climate resilient character of his farm. A person who is energy prodigal at his family level, is expected to go energy extravagant at his farm also.
- Beta coefficient of the causal variable functional education is also negative but significant, it implies that functional education helps to build up consciousness about ecological behavior of farm, its impact on global warming and climate change through different training and sensitization programmes attained by the respondents has been instrumental in farming state or perception on ecological inputs on agriculture and its management.
- The r^2 value stands at 84.00 per cent. It implies that combination of 23 causal variables has explained 84.00 per cent of variance in the consequent variable crop residue left in the field.

Table 5.16: Stepwise Regression Analysis: Climate resilient technology (y_4) Vs. 23 Causal Variables

Sl. No	Variables	Reg.coef. B	S.E. B	Beta	t value
1	Consumption of coal, firewood, fuelwood (kg/year) (x_{22})	0.000	0.000	0.441	3.163
2	Production of organic manure (kg/year) (x_{23})	0.000	0.000	-0.556	-3.857
3	Total marketed surplus of crops cultivated (kg) (x_{12})	2.574E-05	0.000	0.438	3.036
4	Total yield of crops cultivated (kg) (x_{11})	-2.765E-05	0.000	-0.308	-2.147

R square: 53.30%

The standard error of the estimate: 0.632

Model 6.8: Stepwise Regression Analysis: Climate resilient technology (y_4) Vs. 23 Causal Variables

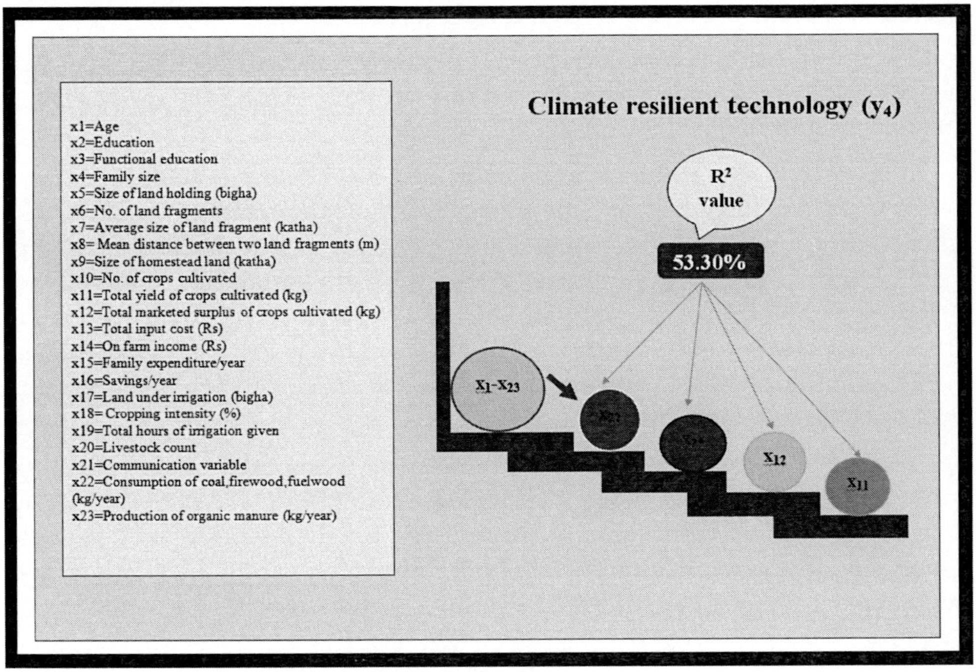

Result: Table 5.16 presents the stepwise regression analysis which elicits that four causal variables, consumption of coal, firewood, fuelwood (kg/year) (x_{22}), production of organic manure (kg/year) (x_{23}), total marketed surplus of crops cultivated (kg) (x_{12}), total yield of crops cultivated (kg) (x_{11}) has come out with stronger determining character on the consequent variable, climate resilient technology (y_4).

Revelation: Table 5.16 presents the stepwise regression analysis for screening out the four causal variables with the most significant functional contribution to the consequent variable climate resilient technology. It has been found that the following variables has been retained at the last step and these are consumption of coal, firewood, fuelwood, production of organic manure, total marketed surplus of crops cultivated, total yield of crops cultivated. The r_2 value stands at 53.30 per cent which implies that these four variables collectively contributed 53.30 per cent of variance in the consequent variable crop residue left in the field. The rest twenty variables have contributed only (84.00-53.30) per cent i.e., 30.70 per cent variance.

The fact has elicited from this hard evidence speaks that variable consumption of coal, firewood, fuelwood has come out as an important determinant for characterizing the climate resilient technology. The energy consumption behavior and pattern of the respondent has got substantive effect on the climate resilient character of his farm. A person who is energy prodigal at his family level, is expected to go energy extravagant at his farm also.

Production of organic manure has come out as an important determinant for characterizing the climate resilient technology. Production of organic manure has been so important in replacing chemical fertilizer, regaining ecological resilience and re-engineering the production process perfectly tuned to ecological functions. When production process goes in compliance with ecological principle and prerequisite, the whole of the production system is certain to go climate resilient.

Total marketed surplus of crops cultivated has come out as an important determinant for characterizing the climate resilient technology. It elicits that whenever the farm keeps generating higher marketed surplus with an increasing trend which implies the farm well-being through sustainable management including ecological issues and inputs. A climate resilient farm means its production pattern will be steady, consistent and upward and that is how higher marketed surplus has been a strong determinant to decide on the climate resilient level and nature of the farm.

Total yield of crops cultivated has come out as an important determinant for characterizing the climate resilient technology. It implies that a climate resilient farm means a farm with higher ecological resilience and function. Whenever application of organic manure, mixing of crop residue, appropriate water and nutrient management, the yield will be positively and effectively impacted.

Table 5.17: Path Analysis: Decomposition of Total Effect into Direct, Indirect and Residual Effect: Climate resilient technology (y_4)

Sl. No	Variables	Total Effect	Direct Effect	Indirect Effect	Highest Indirect Effect
1	Age (x_1)	-0.137	-0.573	0.436	-0.671(x_{17})
2	Education (x_2)	-0.181	0.647	-0.828	0.457(x_{17})
3	Functional education (x_3)	0.153	-0.996	1.149	0.843(x_{17})

Sl. No	Variables	Total Effect	Direct Effect	Indirect Effect	Highest Indirect Effect
4	Family size (x_4)	-0.171	-0.265	0.094	0.956(x_{17})
5	Size of land holding (bigha) (x_5)	-0.083	-0.299	0.216	0.036(x_{17})
6	No. of land fragments (x_6)	0.167	-0.092	0.259	0.113(x_{17})
7	Average size of land fragment (katha) (x_7)	-0.253	**-1.275**	1.022	0.682(x_{17})
8	Mean distance between two land fragments (m) (x_8)	-0.246	0.607	-0.853	0.404(x_{17})
9	Size of homestead land (katha) (x_9)	-0.189	-0.277	0.088	0.237(x_{17})
10	No. of crops cultivated (x_{10})	0.101	-0.499	0.600	0.204(x_{17})
11	Total yield of crops cultivated (kg) (x_{11})	0.074	0.020	0.054	-0.993(x_{17})
12	Total marketed surplus of crops cultivated (kg) (x_{12})	0.090	0.264	-0.174	-0.249(x_{17})
13	Total input cost (Rs) (x_{13})	0.027	-0.330	0.357	0.273(x_9)
14	On farm income (Rs) (x_{14})	-0.019	0.038	-0.057	0.184(x_{17})
15	Family expenditure/year (x_{15})	-0.205	-0.365	0.160	0.918(x_{17})
16	Savings/year (x_{16})	0.148	0.772	-0.624	0.368(x_{17})
17	Land under irrigation (bigha) (x_{17})	0.031	0.258	-0.227	-0.306(x_5)
18	Cropping intensity (%) (x_{18})	-0.064	-0.461	0.397	-0.622(x_{17})
19	Total hours of irrigation given (x_{19})	-0.151	0.283	-0.434	0.692(x_{17})
20	Livestock count (x_{20})	-0.250	0.204	-0.454	-0.193(x_{17})
21	Communication variable (x_{21})	0.004	0.649	-0.645	0.494(x_{17})
22	Consumption of coal,firewood,fuelwood (kg/ year) (x_{22})	0.387	-1.270	**1.657**	-0.259(x_{17})
23	Production of organic manure (kg/year) (x_{23})	-0.359	-0.571	0.212	-0.509(x_{17})

Residual effect: 0.157

Result: Table 5.17 presents the path analysis wherein the total effect of exogenous variable on consequent variable has been decomposed into direct, indirect and residual effect.

Revelation: The causal variable average size of land has exerted the highest direct effect on climate resilient technology and the beta value indicates that the more is the number of fragmentation, the less it will be climate resilient. The increasing scale of fragmentation of small and marginal holding of India with special reference to West Bengal condition indicates that the higher fragmentation has added energy prodigal of the holding. As has already been discussed the faster and intense fragmentation of marginal holding is making the entire farm operation not only energy and cost intensive but also economically non-remunerative.

The causal variable consumption of coal, firewood, fuelwood has exerted the highest indirect effect on climate resilient technology. It justifies that the basic proposition that farmers' energy consumption behavior has also been reflected in the resilient character of his farm under ownership and operation.

The residual value is 0.157 it infers that a little over 15.7 per cent cannot be explain with this combination of 23 variables. It has been supported by the R square value 84.00 per cent as well.

It has been found that the exogenous variable land under irrigation has appeared as many as 21 out of 23 times in exerting the highest effect on the consequent variable climate resilient technology, it implies that in an irrigated agro-based ecosystem the energy and input management system is very much sensitive and we have to be careful because number of irrigation involves energy. When energy is more, then the application of fertilizers will become more. Then again it has got energy equivalent and at the same time when it is intensive agriculture under irrigated agro-based ecosystem, the total number of labour engaged will be more, means it has got energy equivalent. So, that is how it has got tremendous associating effect or ability to influence characters of other variable. For e.g., now the groundwater quality of Punjab and Haryana agriculture was highly intensive and mostly irrigated agro-ecosystem has been there and the entire ecological resilience of Punjab and Haryana agriculture has been seriously depleted and disrupted. The so called green revolution which took place during 1970s, nevertheless it was so sparkling for the phenomenal achievement and at the same time we remain ignorant about how to protect, defend and regenerate our ecological resilience.

Table 5.18: Coefficient of Correlation (r): Soil conservation perception (y_5) and 23 independent variables

Sl. No.	Independent Variables	'r' Value	Remarks
1	Age (x_1)	-0.158	
2	Education (x_2)	0.224	
3	Functional education (x_3)	**0.318**	*
4	Family size (x_4)	-0.084	
5	Size of land holding (bigha) (x_5)	0.178	
6	No. of land fragments (x_6)	-0.103	
7	Average size of land fragment (katha) (x_7)	-0.027	
8	Mean distance between two land fragments (m) (x_8)	0.199	
9	Size of homestead land (katha) (x_9)	0.147	
10	No. of crops cultivated (x_{10})	0.180	

Sl. No.	Independent Variables	'r' Value	Remarks
11	Total yield of crops cultivated (kg) (x_{11})	0.180	
12	Total marketed surplus of crops cultivated (kg) (x_{12})	0.069	
13	Total input cost (Rs) (x_{13})	0.084	
14	On farm income (Rs) (x_{14})	0.207	
15	Family expenditure/year (x_{15})	0.113	
16	Savings/year (x_{16})	-0.114	
17	Land under irrigation (bigha) (x_{17})	0.220	
18	Cropping intensity (%) (x_{18})	0.114	
19	Total hours of irrigation given (x_{19})	0.270	
20	Livestock count (x_{20})	0.010	
21	Communication variable (x_{21})	0.088	
22	Consumption of coal, firewood, fuelwood (kg/year) (x_{22})	0.016	
23	Production of organic manure (kg/year) (x_{23})	-0.011	

**Correlation is significant at the 0.01 level

*Correlation is significant at the 0.05 level

Model 5.9: Coefficient of correlation: Soil conservation perception (y_5) and independent variables having significant correlation.

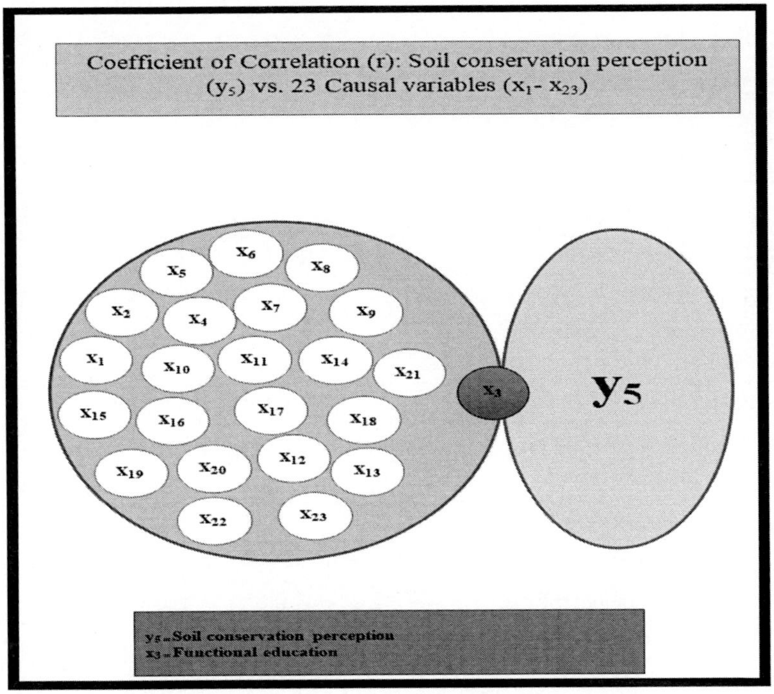

Result: Table 5.18 presents the coefficient of correlation between soil conservation perception (y_5) Vs 23 independent variables (x_1-x_{23}). It has been found that the following variable functional education (x_3) has recorded significant and positive correlation with the dependent variable, soil conservation perception (y_5).

Revelation: Modern propensity of organizing capacity building in the form of training and exposure visits speaks more and more on ecological aspect and off course illusionate implication on economic and technological aspect. Those who are having higher exposure to the training and capacity building program, they have got better perception on soil conservation. The young farmers now have started coming up with an understanding that unless soil erosion are being checked, organic carbon is being restored, microbial populations are enhanced, ecological resilience are retrieved, there will be a serious crisis in agriculture and food security as large.

Table 5.19: Multiple Regression Analysis: Soil conservation perception (y_5) Vs. 23 Causal Variables

Sl. No.	Variables	Reg. Coef. B	S.E. B	Beta	t Value
1	Age (x_1)	-0.007	0.020	-0.122	-0.332
2	Education (x_2)	0.009	0.108	0.028	0.082
3	Functional education (x_3)	-0.271	0.164	**-0.816**	-1.651
4	Family size (x_4)	-0.304	0.118	**-0.772**	-2.580
5	Size of land holding (bigha) (x_5)	0.071	0.087	0.241	0.820
6	No. of land fragments (x_6)	-0.010	0.009	-0.485	-1.046
7	Average size of land fragment (katha) (x_7)	2.694E-06	0.001	0.001	0.002
8	Mean distance between two land fragments (m) (x_8)	0.030	0.030	0.442	0.987
9	Size of homestead land (katha) (x_9)	0.076	0.159	0.214	0.476
10	No. of crops cultivated (x_{10})	-8.003E-06	0.000	-0.124	-0.272
11	Total yield of crops cultivated (kg) (x_{11})	-4.973E-06	0.000	-0.117	-0.344
12	Total marketed surplus of crops cultivated (kg) (x_{12})	8.700E-07	0.000	0.098	0.366
13	Total input cost (Rs) (x_{13})	7.580E-07	0.000	0.359	0.926
14	On farm income (Rs) (x_{14})	3.326E-06	0.000	0.249	0.753
15	Family expenditure/year (x_{15})	1.154E-06	0.000	0.087	0.296
16	Savings/year (x_{16})	0.106	0.086	0.684	1.235
17	Land under irrigation (bigha) (x_{17})	0.001	0.003	0.144	0.359
18	Cropping intensity (%) (x_{18})	0.001	0.001	**0.688**	1.927
19	Total hours of irrigation given (x_{19})	-0.035	0.090	-0.132	-0.388
20	Livestock count (x_{20})	-0.002	0.007	-0.140	-0.360
21	Communication variable (x_{21})	-3.021E-05	0.000	-0.039	-0.157

Sl. No.	Variables	Reg. Coef. B	S.E. B	Beta	t Value
22	Consumption of coal,firewood,fuelwood (kg/year) (x_{22})	-1.598E-05	0.000	-0.056	-0.126
23	Production of organic manure (kg/year) (x_{23})	-0.007	0.020	-0.122	-0.332

R square: 81.60%

The standard error of the estimate: 0.540

Result: Table-5.19 presents the full model of multiple regression analysis between 23 causal variables (x_1-x_{23}) and soil conservation perception (y_5).

Revelation: The table presents multiple regression analysis (full model) between 23 causal variables (x_1-x_{23}) and soil conservation perception (y_5). The three substantives impacts (Beta value) on consequent variable climate resilient technology are in order.

Highlights of the result

Beta coefficient of the causal variable functional education is negative but significant, it implies that most of the training are not purported to the ecological aspect that is why in such cases the functional education mostly hovering around the technological and input aspect so in that case we found discernible impact.

Beta coefficient of the causal variable family size is negative but significant, it implies that where the family size is small the farmer can enjoy a kind of relaxation in terms of economic sustenance rather they can focus more on the ecological aspect of the farm.

Beta coefficient of the causal variable cropping intensity is positive but significant, it implies that whenever the cropping intensity is high, the possibility of soil erosion will also be high. But recently found that the farmers are following different mulching practices to check soil erosion, retain soil moisture and ultimately improve the ecological resilience of farming.

The r^2 value being a combination of all these 23 variables has contributed 81.60 per cent of variance embedded with the consequent variable soil conservation perception.

Table 5.20: Stepwise Regression Analysis: Soil conservation perception (y_5) Vs. 23 Causal Variables

Sl. No	Variables	Reg.coef. B	S.E. B	Beta	t value
1	Total hours of irrigation given (x_{19})	0.001	0.000	0.587	4.102
2	On farm income (Rs) (x_{14})	6.832E-07	0.000	0.324	2.263

R square: 44.70%

The standard error of the estimate: 0.476

Model 5.10: Stepwise Regression Analysis: Soil conservation perception (y_5) Vs. 23 Causal Variables

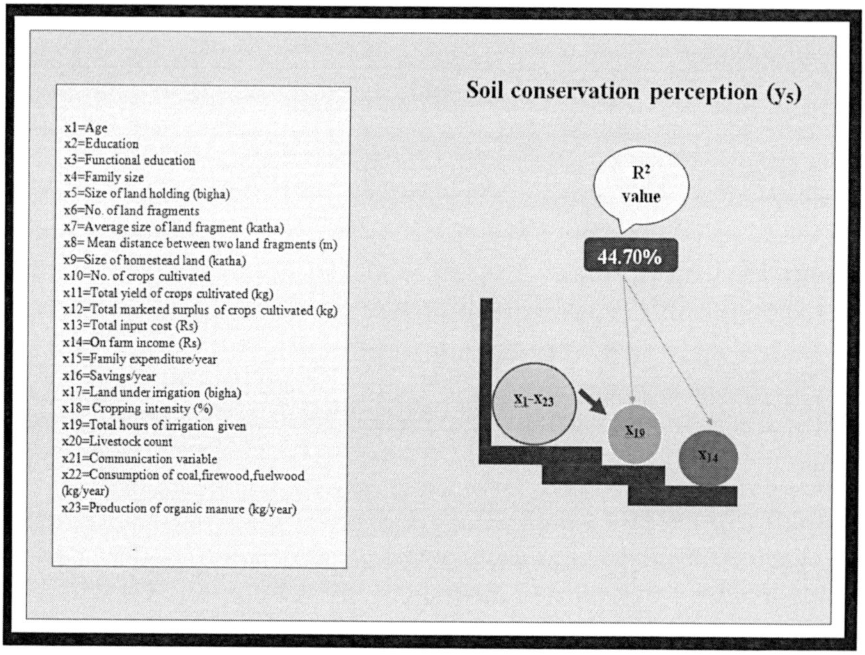

Result: Table 5.20 presents the stepwise regression analysis which elicits that two causal variables,total hours of irrigation given (x_{19}) and on farm income (Rs) (x_{14}) has come out with stronger determining character on the consequent variable, soil conservation perception (y_5).

Revelation: Table 5.20 presents the stepwise regression analysis for screening out the two causal variables with the most significant functional contribution to the consequent variable soil conservation perception. It has been found that the following variables has been retained at the last step and these are total hours of irrigation given and on farm income. The r_2 value stands at 44.70 per cent which implies that these two variables collectively contributed 44.70 per cent of variance in the consequent variable soil conservation perception. The rest twenty variables have contributed only (81.60-44.70) per cent i.e., 36.90 per cent variance.

The fact has elicited from this hard evidence speaks that the natural intensity and approach to irrigation and water management has strongly correlated with erosion of the soil. The farmers need to be sensitive and literate for the parsimonious conjunctive use of irrigation water. So that soil erosion can be reduced, moisture retention can be enhanced and ultimately underground water can be reserved. During field practical observation, it has been observed that through irrigation a channel, the water was not transparent it was turbid. The turbidity of the water indicates that huge amount of top soil here been eroded, the fine particles of soil have been eroded and getting mixed with irrigation water and the water has been opaque and turbid it means due to wrong approach of technology of irrigation water management huge loses of top soil are there from the agricultural field. As we know that in Indian agriculture from per hectare of agricultural land per year around 15 tonnes of top soil are eroded.

Whatever we may suggest, what else we may do, the ultimate decision is taken by the farmers in terms of income. On farm income is the strongest and indomitable dictator on the decision making process of the farmers, this is a reality may be clean, optimistic or may be harsh.

Table 5.21: Path Analysis: Decomposition of Total Effect into Direct, Indirect and Residual Effect: Soil conservation perception (y_5)

Sl. No	Variables	Total Effect	Direct Effect	Indirect Effect	Highest Indirect Effect
1	Age (x_1)	-0.158	-0.122	-0.036	-0.174(x_5)
2	Education (x_2)	0.224	0.028	0.196	0.065(x_5)
3	Functional education (x_3)	0.318	**-0.815**	**1.133**	0.228(x_5)
4	Family size (x_4)	-0.084	-0.772	0.688	0.146(x_5)
5	Size of land holding (bigha) (x_5)	0.178	0.241	-0.063	-0.132(x_{17})
6	No. of land fragments (x_6)	-0.103	-0.484	0.381	0.966(x_5)
7	Average size of land fragment (katha) (x_7)	-0.027	0.001	-0.028	0.997(x_5)
8	Mean distance between two land fragments (m) (x_8)	0.199	0.442	-0.243	0.227(x_5)
9	Size of homestead land (katha) (x_9)	0.147	0.214	-0.067	0.544(x_5)
10	No. of crops cultivated (x_{10})	0.180	-0.124	0.304	0.287(x_5)
11	Total yield of crops cultivated (kg) (x_{11})	0.180	-0.117	0.297	-0.798(x_5)
12	Total marketed surplus of crops cultivated (kg) (x_{12})	0.069	0.096	-0.027	-0.157(x_5)
13	Total input cost (Rs) (x_{13})	0.084	0.359	-0.275	0.354(x_9)
14	On farm income (Rs) (x_{14})	0.207	0.249	-0.042	0.682(x_5)
15	Family expenditure/year (x_{15})	0.113	0.087	0.026	0.525(x_5)
16	Savings/year (x_{16})	-0.114	0.684	-0.798	0.325(x_5)
17	Land under irrigation (bigha) (x_{17})	0.220	0.142	0.078	0.918(x_5)
18	Cropping intensity (%) (x_{18})	0.114	0.688	-0.574	-0.869(x_5)
19	Total hours of irrigation given (x_{19})	0.270	-0.132	0.402	0.827(x_5)
20	Livestock count (x_{20})	0.010	-0.140	0.150	-0.698(x_5)
21	Communication variable (x_{21})	0.088	-0.039	0.127	0.718(x_5)

Sl. No	Variables	Total Effect	Direct Effect	Indirect Effect	Highest Indirect Effect
22	Consumption of coal,firewood,fuelwood (kg/year) (x_{22})	0.016	-0.055	0.071	-0.956(x_5)
23	Production of organic manure (kg/year) (x_{23})	-0.011	-0.120	0.109	-0.951(x_5)

Residual effect: 0.168

Result: Table 5.21 presents the path analysis wherein the total effect of exogenous variable on consequent variable has been decomposed into direct, indirect and residual effect.

Revelation: The causal variable functional education has exerted both the highest direct effect and highest indirect effect on soil conservation perception. The highest direct effect of functional education indicates that the modern propensity of organizing capacity building in the form of training and exposure visits speaks more and more on ecological aspect and off course illusionate implication on economic and technological aspect. Those who are having higher exposure to the training and capacity building program, they have got better perception on soil conservation. The young farmers now have started coming up with an understanding that unless soil erosion are being checked, organic carbon is being restored, microbial populations are enhanced, ecological resilience are retrieved, there will be a serious crisis in agriculture and food security as large.

The indirect effect of functional education has been found to exert highest positive effect but its counterpart that is direct effect has been negative, the reason is that we use to impart training to the farmers but unfortunately the input and other services are not reaching timely to the farmers already trained, that is why the effect of training in many cases are negative or very little.

The residual value is 0.157 it infers that a little over 15.7 per cent cannot be explain with this combination of 23 variables. It has been supported by the R square value 84.00 per cent as well.

It has been found that the exogenous variable size of land holding has appeared as many as 21 out of 23 times in exerting the highest effect on the consequent variable soil conservation perception. Whenever the size of farm is big the farmers' percepts that it is advantageous to leave the more stubble or crop residue for mulching purposes on the field because it is difficult to manage the huge volume of straw and it also curb labour cost.

Table 5.22: Coefficient of Correlation (r): Water conservation perception (y_6) and 23 independent variables

Sl. No.	Independent Variables	'r' Value	Remarks
1	Age (x_1)	-0.156	
2	Education (x_2)	**0.336**	*
3	Functional education (x_3)	0.176	
4	Family size (x_4)	**0.331**	*
5	Size of land holding (bigha) (x_5)	0.258	
6	No. of land fragments (x_6)	0.071	
7	Average size of land fragment (katha) (x_7)	0.061	
8	Mean distance between two land fragments (m) (x_8)	0.276	
9	Size of homestead land (katha) (x_9)	0.162	
10	No. of crops cultivated (x_{10})	**0.320**	*
11	Total yield of crops cultivated (kg) (x_{11})	**0.283**	*
12	Total marketed surplus of crops cultivated (kg) (x_{12})	-0.005	
13	Total input cost (Rs) (x_{13})	0.188	
14	On farm income (Rs) (x_{14})	0.084	
15	Family expenditure/year (x_{15})	0.250	
16	Savings/year (x_{16})	-0.184	
17	Land under irrigation (bigha) (x_{17})	**0.363**	**
18	Cropping intensity (%) (x_{18})	0.187	
19	Total hours of irrigation given (x_{19})	0.076	
20	Livestock count (x_{20})	-0.095	
21	Communication variable (x_{21})	-0.023	
22	Consumption of coal,firewood,fuelwood (kg/year) (x_{22})	-0.247	
23	Production of organic manure (kg/year) (x_{23})	-0.015	

**Correlation is significant at the 0.01 level

*Correlation is significant at the 0.05 level

Result: Table 5.22 presents the coefficient of correlation between water conservation perception (y_6) Vs 23 independent variables (x_1-x_{23}). It has been found that the following variableeducation (x_2), family size (x_4), no. of crops cultivated (x_{10}), total yield of crops cultivated (kg) (x_{11}) and land under irrigation (bigha) (x_{17}) has recorded significant and positive correlation with the dependent variable, water conservation perception (y_6).

Revelation: Education generates ecological learning and that is why where has been better education there has been better water management. Every literate and trained mind nowadays can feel how important the water is. The area where the research has been carried out is falling under new alluvial zone. The main problem of this zone is arsenic contamination in groundwater, not the availability of water as such.

Model 5.11: Coefficient of correlation: Water conservation perception (y_6)and independent variables having significant correlation.

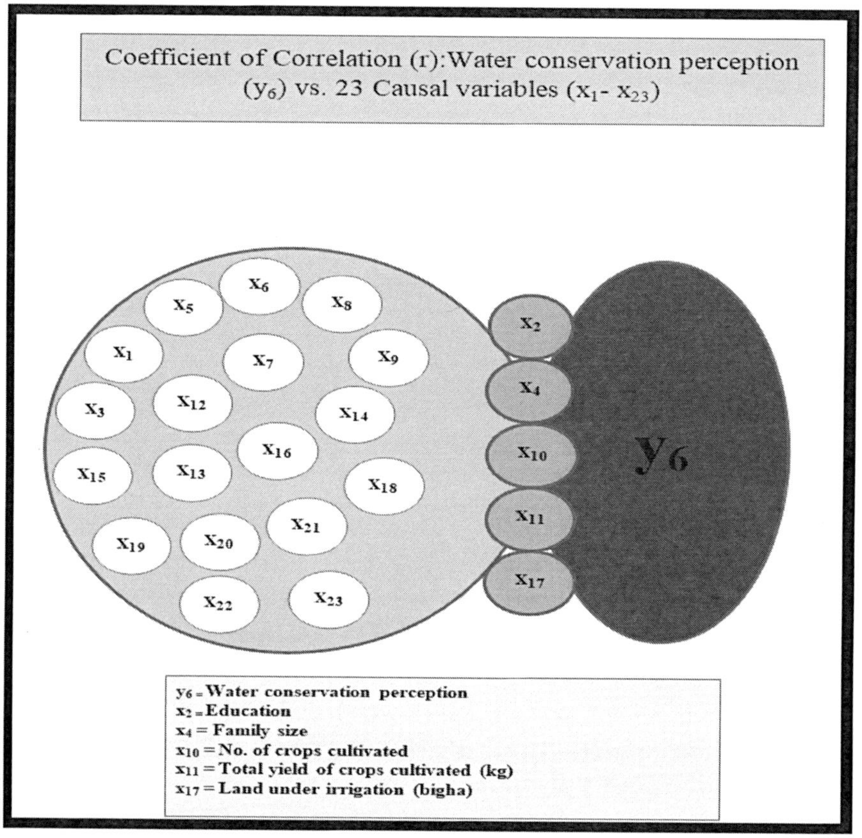

Family size forms and contribute economic, educational and ecological foundation. The results show that for the respondent having higher family size, the water conservation perception has gone up.

The research locale is under irrigated agro-ecosystem, naturally the cropping intensity is high, the higher is the cropping intensity, the more is the count of crops per unit area and hence the water requirement and perception has gone up for the respondent.

Here, the water is monitored, collected and disbursed through an easy access to groundwater, which is plenty, that is how both the yield and cropping intensity has been higher for this research locale. The crop diversity has also been impacted by the easy access to groundwater.

The higher is the size of land under irrigation, the higher will be the water consumption.The intensive and indiscriminate consumption of groundwater has already contributed to arsenic contamination and deterioration of water quality, that is how the farmers in this area following intensive consumption of groundwater has earned an experiential learning towards consumption of water through conjunctive use of irrigation water. The same water applied to the soil can be used for raising another crops and at the same time extending support to the livestock raising.

Table 5.23: Multiple Regression Analysis: Water conservation perception (y_6) and 23 independent variables

Sl. No.	Variables	Reg. Coef. B	S.E. B	Beta	t Value
1	Age (x_1)	0.048	0.035	**0.685**	1.385
2	Education (x_2)	-0.165	0.187	-0.400	-0.881
3	Functional education (x_3)	0.354	0.283	**0.830**	1.251
4	Family size (x_4)	0.236	0.204	0.466	1.161
5	Size of land holding (bigha) (x_5)	0.034	0.150	0.091	0.229
6	No. of land fragments (x_6)	0.006	0.016	0.237	0.381
7	Average size of land fragment (katha) (x_7)	0.003	0.002	**1.057**	1.334
8	Mean distance between two land fragments (m) (x_8)	-0.044	0.052	-0.508	-0.846
9	Size of homestead land (katha) (x_9)	0.296	0.274	0.651	1.078
10	No. of crops cultivated (x_{10})	2.475E-05	0.000	0.298	0.488
11	Total yield of crops cultivated (kg) (x_{11})	1.065E-05	0.000	0.195	0.426
12	Total marketed surplus of crops cultivated (kg) (x_{12})	-5.682E-07	0.000	-0.050	-0.138
13	Total input cost (Rs) (x_{13})	9.044E-07	0.000	0.333	0.639
14	On farm income (Rs) (x_{14})	-9.882E-07	0.000	-0.057	-0.130
15	Family expenditure/year (x_{15})	3.294E-06	0.000	0.192	0.490
16	Savings/year (x_{16})	-0.129	0.148	-0.647	-0.870
17	Land under irrigation (bigha) (x_{17})	-0.005	0.005	-0.554	-1.029
18	Cropping intensity (%) (x_{18})	0.001	0.001	0.505	1.054
19	Total hours of irrigation given (x_{19})	-0.120	0.155	-0.352	-0.770
20	Livestock count (x_{20})	-0.009	0.012	-0.424	-0.811
21	Communication variable (x_{21})	0.000	0.000	-0.297	-0.897
22	Consumption of coal,firewood,fuelwood (kg/year) (x_{22})	0.000	0.000	0.360	0.602
23	Production of organic manure (kg/year) (x_{23})	0.048	0.035	**0.685**	1.385

R square: 66.90% The standard error of the estimate: 0.932

Result: Table-5.23 presents the full model of multiple regression analysis between 23 causal variables (x_1-x_{23}) and water conservation perception (y_6).

Revelation: The table presents multiple regression analysis (full model) between 23 causal variables (x_1-x_{23}) and water conservation perception (y_6). The three substantives impact (Beta value) on consequent variable water conservation perception are in order.

Highlights of the result

- Beta coefficient of the causal variable average size of land fragment is positive and significant, it implies that the water management and use efficiency depends on the size and fragmentation of the land holding. When the size of land holding is small and fragmentations are more, it is quite likely that water use efficiency will be negatively affected. More water will go after waste and cost of irrigation will go up as well.
- Beta coefficient of the causal variable functional education is positive and significant, it implies that the modern propensity of organizing capacity building in the form of training and exposure visits speaks more and more on ecological aspect and off course illusionate implication on economic and technological aspect. Those who are having higher exposure to the training and capacity building program, they have got better perception on water conservation. The young farmers now have started coming up with an understanding that water conservation helps life saving irrigation during critical drought period,destruction of natural water bodies amounts to loss of plant and fish biodiversity, water conservation must focus on judicious use of groundwater, of more on rainwater harvesting, or else there will be a serious crisis in agriculture and food security as large.
- Both the causal variable age and production of organic manure have recorded same beta coefficient value, both the value are positive and significant. Chronological age means level of maturity and experience, the higher is the age, better has been the perception of water conservation. Farmers are increasingly using organic manure to improve the soil quality, both physically and biologically. Enhanced use of organic manure rationalizes the consumption of irrigation water.
- The r^2 value being a combination of all these 23 variables has contributed 66.90 per cent of variance embedded with the consequent variable water conservation perception.

Table 5.24: Stepwise Regression Analysis: Water conservation perception (y_6) and 23 independent variables

Sl. No	Variables	Reg.coef. B	S.E. B	Beta	t value
1	No. of crops cultivated (x_{10})	0.209	0.069	0.460	3.030
2	Production of organic manure (kg/ year) (x_{23})	0.000	0.000	-0.395	-2.635
3	Average size of land fragment (katha) (x_7)	0.009	0.004	0.342	2.254

R square: 41.80%

The standard error of the estimate: 0.641

Model 5.12: Stepwise Regression Analysis: Water conservation perception (y_6) Vs. 23 Causal Variables

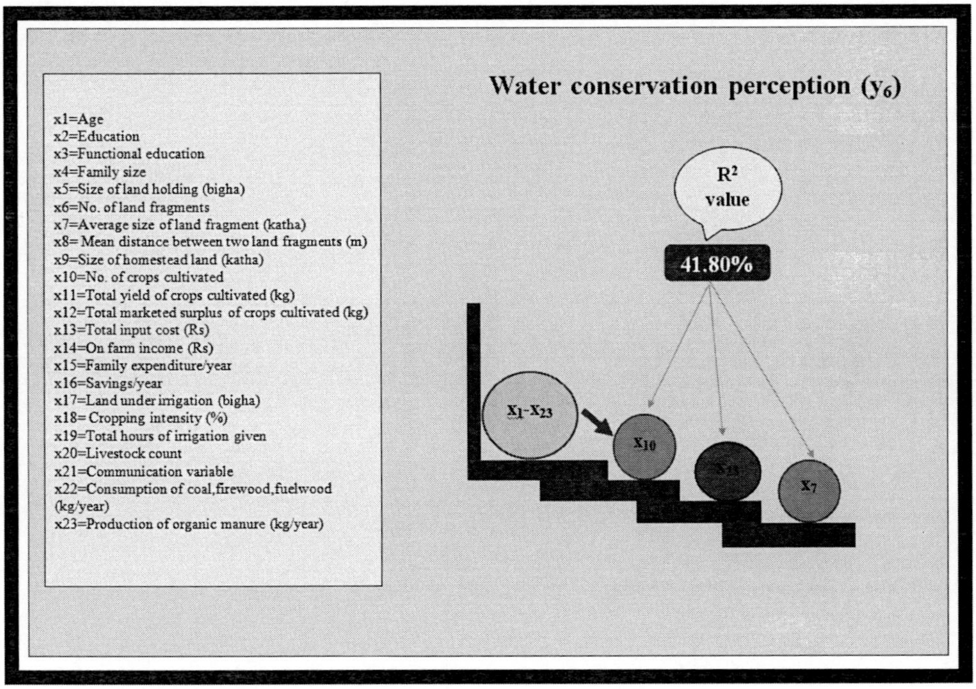

Result: Table 5.24 presents the stepwise regression analysis which elicits that three causal variables,no. of crops cultivated (x_{10}), production of organic manure (kg/year) (x_{23}) and average size of land fragment (katha) (x_7) has come out with stronger determining character on the consequent variable, water conservation perception (y_6).

Revelation: Table 5.24 presents the stepwise regression analysis for screening out the three causal variables with the most significant functional contribution to the consequent variable water conservation perception. It has been found that the following variables have been retained at the last step and these are no. of crops cultivated, production of organic manure and average size of land fragment The r^2 value stands at 41.80 per cent which implies that these three variables collectively contributed 41.80 per cent of variance in the consequent variable water conservation perception. The rest twenty variables have contributed only [(41.80/66.90) *100] per cent i.e., 62.48per cent variance.

The fact has elicited from this hard evidence speaks that the research locale is under irrigated agro-ecosystem, naturally the cropping intensity is high, the higher is the cropping intensity, the more is the count of crops per unit area and hence the water requirement and perception has gone up for the respondent. Farmers are increasingly using organic manure to improve the soil quality, both physically and biologically. Enhanced use of organic manure rationalizes the consumption of irrigation water. the water management and use efficiency depends on the size and fragmentation of the land holding. When the size of land holding is small and fragmentations are more, it is quite likely that water use efficiency will be negatively affected. More water will go after waste and cost of irrigation will go up as well. .

Table 5.25: Path Analysis: Decomposition of Total Effect into Direct, Indirect and Residual Effect: Water conservation perception (y_6) and 23 independent variables

Sl. No	Variables	Total Effect	Direct Effect	Indirect Effect	Highest Indirect Effect
1	Age (x_1)	-0.156	0.685	-0.841	0.991(x_{17})
2	Education (x_2)	0.336	-0.4	0.736	-0.738(x_{17})
3	Functional education (x_3)	0.176	0.829	-0.653	-0.668(x_{17})
4	Family size (x_4)	0.331	0.466	-0.135	-0.909(x_{17})
5	Size of land holding (bigha) (x_5)	0.258	0.091	0.167	-0.946(x_{17})
6	No. of land fragments (x_6)	0.071	0.237	-0.166	0.292(x_8)
7	Average size of land fragment (katha) (x_7)	0.061	**1.057**	**-0.996**	-0.220(x_{17})
8	Mean distance between two land fragments (m) (x_8)	0.276	-0.508	0.784	-0.074(x_{17})
9	Size of homestead land (katha) (x_9)	0.162	0.651	-0.489	-0.370(x_{17})
10	No. of crops cultivated (x_{10})	0.32	0.298	0.022	-0.296(x_{18})
11	Total yield of crops cultivated (kg) (x_{11})	0.283	0.194	0.089	-0.280(x_8)
12	Total marketed surplus of crops cultivated (kg) (x_{12})	-0.005	-0.05	0.045	-0.292(x_8)
13	Total input cost (Rs) (x_{13})	0.188	0.333	-0.145	-0.277(x_9)
14	On farm income (Rs) (x_{14})	0.084	-0.057	0.141	-0.707(x_{17})

Sl. No	Variables	Total Effect	Direct Effect	Indirect Effect	Highest Indirect Effect
15	Family expenditure/year (x_{15})	0.25	0.192	0.058	-0.448(x_{17})
16	Savings/year (x_{16})	-0.184	-0.647	0.463	-0.271(x_{17})
17	Land under irrigation (bigha) (x_{17})	0.363	-0.554	0.917	0.296(x_5)
18	Cropping intensity (%) (x_{18})	0.187	0.504	-0.317	0.642(x_{17})
19	Total hours of irrigation given (x_{19})	0.076	-0.352	0.428	-0.308(x_{17})
20	Livestock count (x_{20})	-0.095	-0.422	0.327	-0.298(x_3)
21	Communication variable (x_{21})	-0.023	-0.297	0.274	-0.226(x_1)
22	Consumption of coal,firewood,fuelwood (kg/year) (x_{22})	-0.247	0.36	-0.607	0.487(x_{17})
23	Production of organic manure (kg/year) (x_{23})	-0.015	0.684	-0.699	0.401(x_{17})

Residual effect: 0.321

Result: Table 5.25 presents the path analysis wherein the total effect of exogenous variable on consequent variable water conservation perception has been decomposed into direct, indirect and residual effect.

Revelation: The causal variable average size of land fragment has exerted both the highest direct effect and highest indirect effect on water conservation perception. The highest direct effect of average size of land fragment indicates that the water management and use efficiency depends on the size and fragmentation of the land holding. When the size of land holding is small and fragmentations are more, it is quite likely that water use efficiency will be negatively affected. More water will go after waste and cost of irrigation will go up as well.

The indirect effect of average size of land holding has been found to exert highest negative effect on the consequent variable water conservation perception, the reason is that the average size of land holding denotes the size of land resources. The higher is the average size of land holding, the higher would be the water consumption. The monitoring and management of water consumption here are being dictated by the size of land resources at large. The impact of average size of land holding on water conservation perception also impacted by the number of fragmentation held by the given size of land holding.

The residual value is 0.321 it infers that a little over 32.1 per cent cannot be explain with this combination of 23 variables. It has been supported by the R square value 66.90 per cent as well.

It has been found that the exogenous variable land under irrigation has appeared as many as 21 out of 23 times in exerting the highest effect on the consequent variable water conservation perception, it implies that in an irrigated agro-based ecosystem the energy and input management system is very much sensitive and we have to be careful because number of irrigation involves energy. When energy is more, then the application of fertilizers will become more. Then again it has got energy equivalent and at the same time when it is intensive agriculture under irrigated agro-based ecosystem, the total number of labour engaged will be more, means it has got energy equivalent. So, that is how it has got tremendous associating effect or ability to influence characters of other variable.

Table 5.26: Coefficient of Correlation (r): Biodiversity conservation perception(y_7)

Sl. No.	Independent Variables	'r' Value	Remarks
1	Age (x_1)	0.214	
2	Education (x_2)	-0.143	
3	Functional education (x_3)	-0.149	
4	Family size (x_4)	-0.143	
5	Size of land holding (bigha) (x_5)	-0.199	
6	No. of land fragments (x_6)	-0.005	
7	Average size of land fragment (katha) (x_7)	**-0.296**	*
8	Mean distance between two land fragments (m) (x_8)	**-0.343**	*
9	Size of homestead land (katha) (x_9)	0.071	
10	No. of crops cultivated (x_{10})	0.116	
11	Total yield of crops cultivated (kg) (x_{11})	-0.009	
12	Total marketed surplus of crops cultivated (kg) (x_{12})	0.073	
13	Total input cost (Rs) (x_{13})	-0.080	
14	On farm income (Rs) (x_{14})	0.094	
15	Family expenditure/year (x_{15})	-0.220	
16	Savings/year (x_{16})	-0.125	
17	Land under irrigation (bigha) (x_{17})	**-0.304**	*
18	Cropping intensity (%) (x_{18})	0.090	
19	Total hours of irrigation given (x_{19})	0.070	
20	Livestock count (x_{20})	-0.056	
21	Communication variable (x_{21})	-0.231	
22	Consumption of coal,firewood,fuelwood (kg/year) (x_{22})	-0.020	
23	Production of organic manure (kg/year) (x_{23})	-0.069	

**Correlation is significant at the 0.01 level

*Correlation is significant at the 0.05 level

Model 5.13: Coefficient of correlation: Biodiversity conservation perception(y_7)and independent variables having significant correlation.

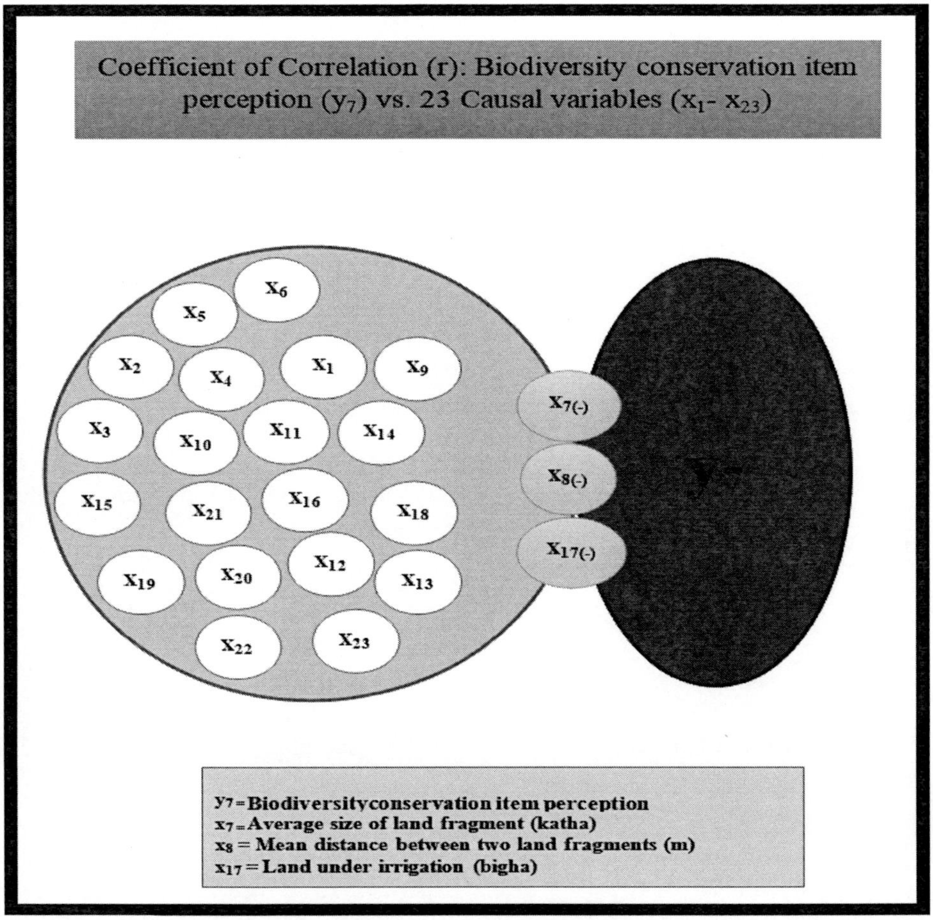

Result: Table 5.26 presents the coefficient of correlation between biodiversity conservation perception (y_7) Vs 23 independent variables (x_1-x_{23}). It has been found that the following variable average size of land fragment (katha) (x_7), mean distance between two land fragments (m) (x_8), land under irrigation (bigha) (x_{17}) has recorded significant and negative correlation with the dependent variable, biodiversity conservation perception (y_7).

Revelation: The variable average size of land fragment has recorded significant and negative correlation with biodiversity conservation perception. This is quite obvious because the more is the fragmentation, the higher is the adverse effect on biodiversity. Land fragmentation does not allow diversity, floral and faunal to grow up and evolve. It is due to tiny size of operating land, there is no scope to have a pond or fish of marshy land along side of the main cropping unit. So the diversity has been negatively impacted with the more number of fragmentations.

The variable mean distance between two land fragments have recorded significant and negative correlation with biodiversity conservation perception. This implies that when land is not consolidated, the fragments are strawling around, the biodiversity will be less.

The variable land under irrigation has recorded significant and negative correlation with biodiversity conservation perception. It is well discernible that irrigation does not promote natural biodiversity because irrigation in agriculture is associated with application of pesticides and heavy doses of chemical fertilizers. Both these are deleterious to the protection and growth of biodiversity. The more area brought under irrigation, means it passes through modernization and mechanization process. This modernization and mechanization process have contributed to huge loses of local cultivar. Green diversity turned skewed and myopic.

Table 5.27: Multiple Regression Analysis: Biodiversity conservation perception(y_7) Vs. 23 Causal Variables

Sl. No.	Variables	Reg. Coef. B	S.E. B	Beta	t Value
1	Age (x_1)	0.009	0.015	0.155	0.633
2	Education (x_2)	-0.012	0.080	-0.033	-0.147
3	Functional education (x_3)	-0.060	0.121	-0.161	-0.491
4	Family size (x_4)	-0.049	0.087	-0.113	-0.567
5	Size of land holding (bigha) (x_5)	0.127	0.064	0.384	1.971
6	No. of land fragments (x_6)	-0.029	0.007	**-1.293**	-4.201
7	Average size of land fragment (katha) (x_7)	0.001	0.001	0.443	1.131
8	Mean distance between two land fragments (m) (x_8)	-0.008	0.022	-0.104	-0.351
9	Size of homestead land (katha) (x_9)	-0.093	0.117	-0.236	-0.792
10	No. of crops cultivated (x_{10})	3.979E-05	0.000	0.552	1.830
11	Total yield of crops cultivated (kg) (x_{11})	8.092E-06	0.000	0.171	0.756
12	Total marketed surplus of crops cultivated (kg) (x_{12})	-3.343E-07	0.000	-0.034	-0.190
13	Total input cost (Rs) (x_{13})	1.265E-06	0.000	0.537	2.087
14	On farm income (Rs) (x_{14})	5.624E-08	0.000	0.004	0.017
15	Family expenditure/year (x_{15})	3.542E-06	0.000	0.238	1.229
16	Savings/year (x_{16})	-0.014	0.063	-0.081	-0.219
17	Land under irrigation (bigha) (x_{17})	0.000	0.002	-0.056	-0.210
18	Cropping intensity (%) (x_{18})	0.001	0.001	**0.659**	2.781

Sl. No.	Variables	Reg. Coef. B	S.E. B	Beta	t Value
19	Total hours of irrigation given (x_{19})	-0.086	0.067	-0.291	-1.288
20	Livestock count (x_{20})	-0.008	0.005	-0.437	-1.693
21	Communication variable (x_{21})	0.000	0.000	-0.389	-2.381
22	Consumption of coal,firewood,fuelwood (kg/year) (x_{22})	0.000	0.000	**0.679**	2.299
23	Production of organic manure (kg/ year) (x_{23})	0.009	0.015	0.155	0.633

R square: 91.90%

The standard error of the estimate: 0.399

Result: Table-5.27 presents the full model of multiple regression analysis between 23 causal variables (x_1-x_{23}) and biodiversity conservation perception (y_7).

Revelation: The table presents multiple regression analysis (full model) between 23 causal variables (x_1-x_{23}) and biodiversity conservation perception (y_7). The three substantives impacts (Beta value) on consequent variable biodiversity conservation perception are in order.

Highlights of the result

- Beta coefficient of the causal variable no. of land fragments is negative and significant on the consequent variable biodiversity conservation perception, it implies that the more is the fragmentation, the higher is the adverse effect on biodiversity. Land fragmentation does not allow diversity, floral and faunal to grow up and evolve. It is due to tiny size of operating land, there is no scope to have a pond or fish of marshy land along side of the main cropping unit. So the diversity has been negatively impacted with the more number of fragmentations.
- Beta coefficient of the causal variableconsumption of coal, firewood, fuelwood is positive and significant on the consequent variable biodiversity conservation perception, it implies thatconsumption of coal, firewood, fuelwood has contributed to reduction of biodiversity and collection of firewood is also associated with grazing practices. The livestock grazers used to collect firewood while their livestock are on grazing.
- Beta coefficient of the causal variable cropping intensity is positive and significant on the consequent variable biodiversity conservation perception, it implies that if cropping intensity is increased due to the repetition of same crop on the same piece of land, as has taken place for rice in West Bengal, the possibility and scope of biodiversity improvement will be reduced. The increase of cropping intensity in West Bengal is attributed by the repetition of

rice not by other crops like pulse and oilseeds. On the other hand, the practice of intercropping with rice is not that visible in West Bengal.

The r^2 value being a combination of all these 23 variables has contributed 91.90 per cent of variance embedded with the consequent variable biodiversity conservation perception.

Table 5.28: Stepwise Regression Analysis: Biodiversity conservation perception(y_7) Vs. 23 Causal Variables

Sl. No	Variables	Reg.coef. B	S.E. B	Beta	t value
1	Average size of land fragment (katha) (x_7)	-0.019	0.003	-0.855	-7.665
2	Total yield of crops cultivated (kg) (x_{11})	1.940E-05	0.000	0.269	2.413

R square: 68.60%

The standard error of the estimate: 0.401

Model 5.14: Stepwise Regression Analysis: Biodiversity conservation perception (y_7 Vs. 23 Causal Variables

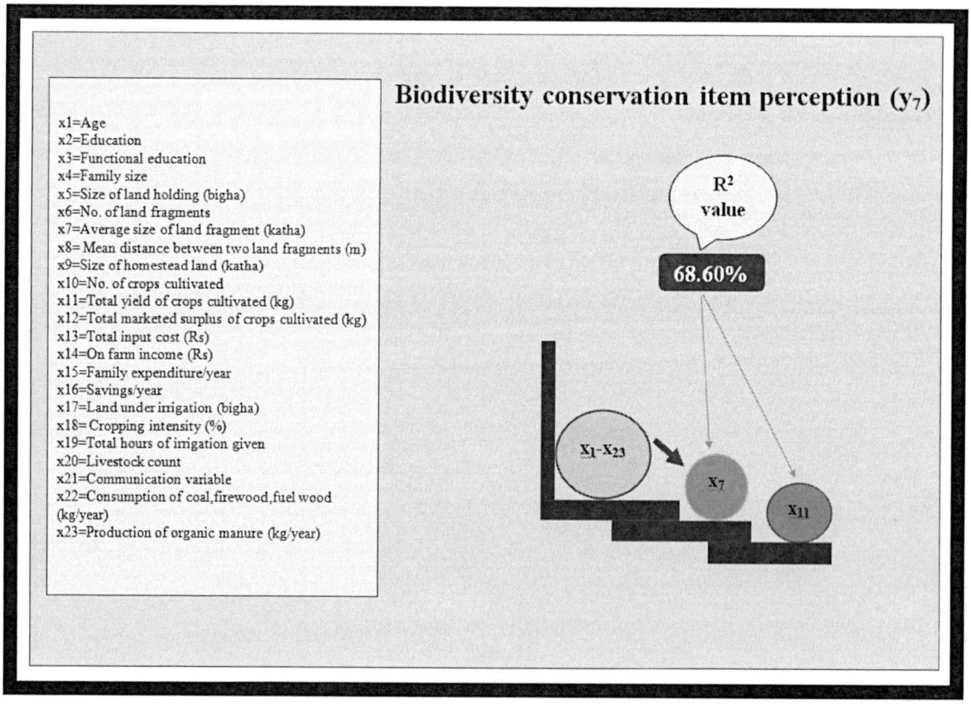

Result: Table 5.28 presents the stepwise regression analysis which elicits that two causal variables, average size of land fragment (katha) (x_7), total yield of crops cultivated (kg) (x_{11}) has come out with stronger determining character on the consequent variable, biodiversity conservation perception (y_7).

Revelation: Table 5.28 presents the stepwise regression analysis for screening out the three causal variables with the most significant functional contribution to the consequent variable biodiversity conservation perception. It has been found that the following variables has been retained at the last step and these are average size of land fragment and total yield of crops cultivated. The r^2 value stands at 68.80 per cent which implies that these two variables collectively contributed 68.80 per cent of variance in the consequent variable biodiversity conservation perception. The rest twenty variables have contributed only [(68.80/91.90) *100] per cent i.e., 74.86per cent variance.

The fact has elicited from this hard evidence speaks that the more is the fragmentation, the higher is the adverse effect on biodiversity. Land fragmentation does not allow diversity, floral and faunal to grow up and evolve. It is due to tiny size of operating land, there is no scope to have a pond or fish of marshy land along side of the main cropping unit. So the diversity has been negatively impacted with the more number of fragmentations.

When the total yield of crops is mostly contributed by a solitary crop or some few crops only, the scope of biodiversity is only limited as has mostly been happened to rice based cropping system in this area.

Table 5.29: Path Analysis: Decomposition of Total Effect into Direct, Indirect and Residual Effect: Biodiversity conservation perception (y7)

Sl. No	Variables	Total Effect	Direct Effect	Indirect Effect	Highest Indirect Effect
1	Age (x_1)	0.214	0.155	0.059	0.036(x_{17})
2	Education (x_2)	-0.143	-0.033	-0.11	-0.592(x_{17})
3	Functional education (x_3)	-0.149	-0.161	0.012	-0.155(x_{17})
4	Family size (x_4)	-0.143	-0.111	-0.032	-0.658(x_{17})
5	Size of land holding (bigha) (x_5)	-0.199	0.383	-0.582	-0.115(x_{17})
6	No. of land fragments (x_6)	-0.005	**-1.293**	**1.288**	-0.502(x_{17})
7	Average size of land fragment (katha) (x_7)	-0.296	0.443	-0.739	-0.596(x_{17})
8	Mean distance between two land fragments (m) (x_8)	-0.343	-0.104	-0.239	-0.687(x_{17})
9	Size of homestead land (katha) (x_9)	0.071	-0.236	0.307	-0.302(x_{17})
10	No. of crops cultivated (x_{10})	0.116	0.551	-0.435	0.219(x_{11})
11	Total yield of crops cultivated (kg) (x_{11})	-0.009	0.171	-0.18	0.417(x_{17})
12	Total marketed surplus of crops cultivated (kg) (x_{12})	0.073	-0.031	0.104	0.599(x_{17})
13	Total input cost (Rs) (x_{13})	-0.08	0.537	-0.617	0.133(x_{17})

Sl. No	Variables	Total Effect	Direct Effect	Indirect Effect	Highest Indirect Effect
14	On farm income (Rs) (x_{14})	0.094	0.004	0.09	-0.398(x_{17})
15	Family expenditure/year (x_{15})	-0.22	0.236	-0.456	0.790(x_5)
16	Savings/year (x_{16})	-0.125	-0.081	-0.044	-0.684(x_{17})
17	Land under irrigation (bigha) (x_{17})	-0.304	-0.056	-0.248	0.077(x_5)
18	Cropping intensity (%) (x_{18})	0.09	0.659	-0.569	0.998(x_{17})
19	Total hours of irrigation given (x_{19})	0.07	-0.291	0.361	0.932(x_5)
20	Livestock count (x_{20})	-0.056	-0.437	0.381	-0.857(x_5)
21	Communication variable (x_{21})	-0.231	-0.388	0.157	0.362(x_5)
22	Consumption of coal,firewood,fuelwood (kg/year) (x_{22})	-0.02	0.679	-0.699	0.029(x_{17})
23	Production of organic manure (kg/year) (x_{23})	-0.069	0.153	-0.222	-0.500(x_5)

Residual effect: 0.079

Result: Table 5.29 presents the path analysis wherein the total effect of exogenous variable on consequent variable biodiversity conservation perception has been decomposed into direct, indirect and residual effect.

Revelation: The causal variable average no. of land fragments has exerted both the highest direct effect and highest indirect effect on biodiversity conservation perception. Faster and more intensive rate of fragmentation has made our farming more and more energy and resource prodigal. It does not allow or sustain higher biodiversity and ecological resilience. When a tiny and fragmented holding is associated with high intensity chemicalization in form of pesticides and fertilizer application, indiscriminate depletion of groundwater, monotonous and myopic varietal selection over the year, the whole of the ecological function and resilience are severely disrupted. The need of the hour is to go for community level management of farm operation planning and monitoring and operation of small holding to make our farming more tune to ecological practices and at the same time higher risk deterrent.

The residual value is 0.079it infers that a little over 7.9 per cent cannot be explain with this combination of 23 variables. It has been supported by the R square value 91.90 per cent as well.

It has been found that the exogenous variable land under irrigation has appeared as many as 16 out of 23 times in exerting the highest effect on the consequent variable biodiversity conservation perception, it implies that in an irrigated agro-based ecosystem the energy and input management system is very much sensitive and we have to be careful because number of irrigation involves energy. When energy is more, then the application of fertilizers will become more. Then again it has got energy equivalent and at the same time when it is intensive agriculture under irrigated agro-based ecosystem, the total number of labour engaged will be more, means it has got energy equivalent. So, that is how it has got tremendous associating effect or ability to influence characters of other variable.

Table 5.30: Coefficient of Correlation (r): Soil stewardship perception(y_8)

Sl. No.	Independent Variables	'r' Value	Remarks
1	Age (x_1)	0.195	
2	Education (x_2)	0.238	
3	Functional education (x_3)	-0.026	
4	Family size (x_4)	0.251	
5	Size of land holding (bigha) (x_5)	0.148	
6	No. of land fragments (x_6)	0.000	
7	Average size of land fragment (katha) (x_7)	0.081	
8	Mean distance between two land fragments (m) (x_8)	-0.103	
9	Size of homestead land (katha) (x_9)	-0.008	
10	No. of crops cultivated (x_{10})	0.234	
11	Total yield of crops cultivated (kg) (x_{11})	**0.392**	**
12	Total marketed surplus of crops cultivated (kg) (x_{12})	-0.183	
13	Total input cost (Rs) (x_{13})	-0.098	
14	On farm income (Rs) (x_{14})	-0.087	
15	Family expenditure/year (x_{15})	0.268	
16	Savings/year (x_{16})	-0.129	
17	Land under irrigation (bigha) (x_{17})	0.148	
18	Cropping intensity (%) (x_{18})	**0.319**	*
19	Total hours of irrigation given (x_{19})	0.055	
20	Livestock count (x_{20})	0.008	
21	Communication variable (x_{21})	-0.110	
22	Consumption of coal,firewood,fuelwood (kg/year) (x_{22})	-0.093	
23	Production of organic manure (kg/year) (x_{23})	-0.149	

**Correlation is significant at the 0.01 level

*Correlation is significant at the 0.05 level

Model 5.15: Coefficient of correlation: Soil stewardship perception (y_8) and independent variables having significant correlation.

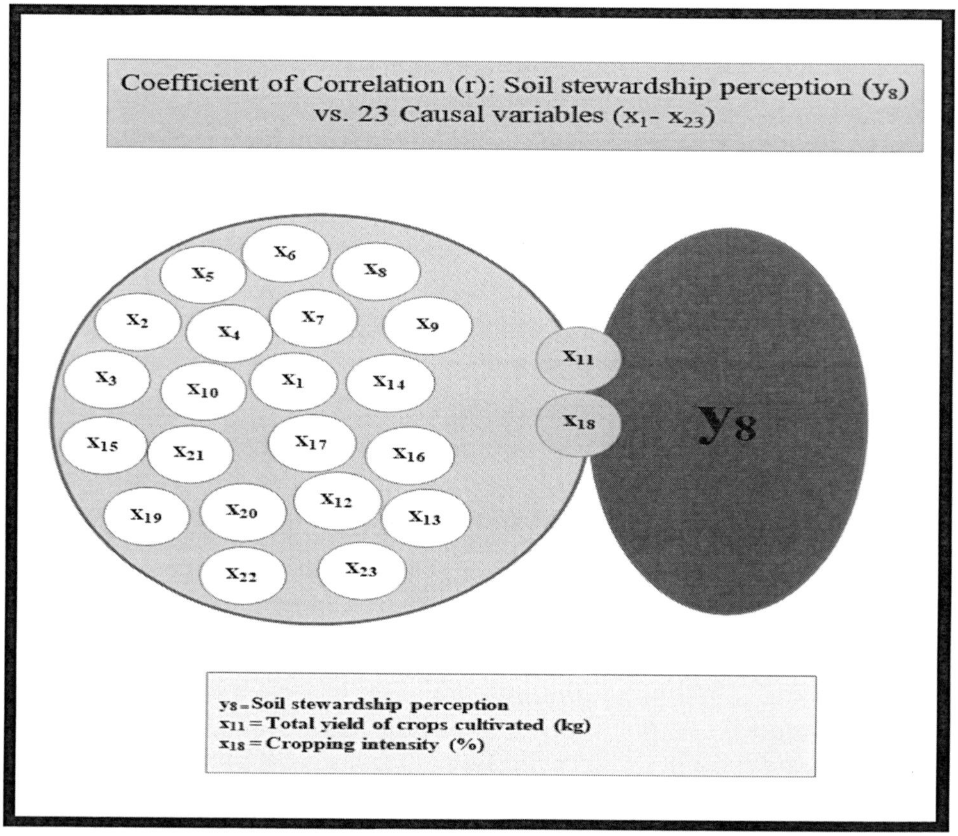

Result: Table 5.30 presents the coefficient of correlation between soil stewardship perception (y_8) Vs 23 independent variables (x_1-x_{23}). It has been found that the following variable total yield of crops cultivated (kg) (x_{11}) and cropping intensity (%) (x_{18}) has recorded significant and positive correlation with the dependent variable, soil stewardship perception (y_8).

Revelation: The variable total yield of crops cultivated and cropping intensity has recorded significant and positive correlation with soil stewardship perception. This is quite obvious because the fertility status and management of the soil has got direct contribution to yield behaviour of the crop. It sustains and fosters cropping intensity that is how, cropping intensity and perception of soil stewardship are interdependent on each other and exert reciprocal bearing on each other as well.

Table 5.31: Multiple Regression Analysis: Soil stewardship perception (y_8) Vs. 23 Causal Variables

Sl. No.	Variables	Reg. Coef. B	S.E. B	Beta	t Value
1	Age (x_1)	-0.012	0.018	-0.218	-0.682
2	Education (x_2)	0.078	0.097	0.236	0.804
3	Functional education (x_3)	-0.149	0.146	-0.437	-1.019
4	Family size (x_4)	0.034	0.105	0.083	0.319
5	Size of land holding (bigha) (x_5)	-0.046	0.077	-0.151	-0.591
6	No. of land fragments (x_6)	0.001	0.008	0.062	0.153
7	Average size of land fragment (katha) (x_7)	-0.002	0.001	**-0.951**	-1.856
8	Mean distance between two land fragments (m) (x_8)	-0.011	0.027	-0.154	-0.397
9	Size of homestead land (katha) (x_9)	0.143	0.141	0.395	1.011
10	No. of crops cultivated (x_{10})	1.520E-05	0.000	0.229	0.581
11	Total yield of crops cultivated (kg) (x_{11})	-1.961E-05	0.000	-0.451	-1.521
12	Total marketed surplus of crops cultivated (kg) (x_{12})	3.220E-06	0.000	0.355	1.520
13	Total input cost (Rs) (x_{13})	-8.160E-07	0.000	-0.377	-1.118
14	On farm income (Rs) (x_{14})	3.439E-06	0.000	0.251	0.874
15	Family expenditure/year (x_{15})	-3.048E-06	0.000	-0.223	-0.878
16	Savings/year (x_{16})	0.103	0.076	**0.649**	1.350
17	Land under irrigation (bigha) (x_{17})	0.001	0.003	0.132	0.379
18	Cropping intensity (%) (x_{18})	-0.001	0.001	**-0.634**	-2.046
19	Total hours of irrigation given (x_{19})	0.134	0.080	0.496	1.678
20	Livestock count (x_{20})	-0.004	0.006	-0.242	-0.718
21	Communication variable (x_{21})	0.000	0.000	0.262	1.226
22	Consumption of coal,firewood,fuelwood (kg/year) (x_{22})	0.000	0.000	-0.592	-1.531
23	Production of organic manure (kg/year) (x_{23})	-0.012	0.018	-0.218	-0.682

R square: 86.10%

The standard error of the estimate: 0.481

Result: Table-5.27 presents the full model of multiple regression analysis between 23 causal variables (x_1-x_{23}) and soil stewardship perception (y_8).

Revelation: The table-5.27 presents multiple regression analysis (full model) between 23 causal variables (x_1-x_{23}) and soil stewardship perception (y_8). The three substantives impacts (Beta value) on consequent variable soil stewardship perception are in order.

Highlights of the result

- Beta coefficient of the causal variable average size of land fragment is negative and significant on the consequent variable soil stewardship perception, it implies that with the increase in number and intensity of fragmentation it becomes difficult to manage the soil and uphold its fertility status towards attaining a sustainable character of agriculture. When fertility status is maintained, through soil stewardship crater by the farmer, it is expected that the ecological resilience of the agro-ecosystem as a whole will be maintained. It has also been observed that the farmers are becoming increasingly conscious about the deterioration of soil health, as reflected through low organic carbon content in the soil and they are increasingly conscious to follow soil stewardship through application of biofertilizers, testing of soil and increased adoption of mulching practices.
- Beta coefficient of the causal variable savings/years positive and significant on the consequent variable soil stewardship perception, it implies that when soil stewardship is maintained in the form of application of organic manure, soil testing, soil mulching, crop residue mixing in the soil, the yield become sustainable and income generation from the unit of land becomes consistent.
- Beta coefficient of the causal variable is negative and significant on the consequent variable soil stewardship perception, it implies that when the soil health is better, cropping intensity will be higher.

The r^2 value being a combination of all these 23 variables has contributed 86.10 per cent of variance embedded with the consequent variable soil stewardship perception.

Table 5.32: Stepwise Regression Analysis: Soil stewardship perception (y_8) Vs. 23 Causal Variables

Sl. No	Variables	Reg.coef. B	S.E. B	Beta	t value
1	Communication variable (x_{21})	-0.008	0.002	-0.464	-3.483
2	Total yield of crops cultivated (kg) (x_{11})	2.903E-05	0.000	0.438	3.204
3	Size of homestead land (katha) (x_9)	-0.023	0.009	-0.333	-2.508
4	Family size (x_4)	0.118	0.056	0.293	2.131

R square: 57.20%

The standard error of the estimate: 0.447

Model 5.16: Stepwise Regression Analysis: Soil stewardship perception (y_8) Vs. 23 Causal Variables

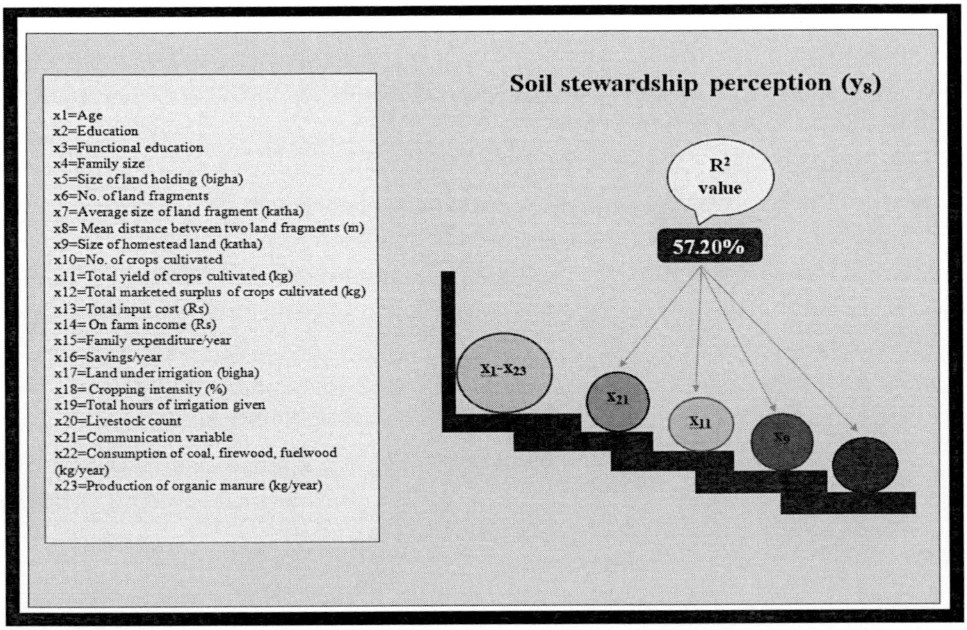

Result: Table 5.32 presents the stepwise regression analysis which elicits that four causal variables, communication variable (x_{21}), total yield of crops cultivated (kg) (x_{11}), size of homestead land (katha) (x_9) and family size (x_4) has come out with stronger determining character on the consequent variable, soil stewardship perception (y_8).

Revelation: Table 5.32 presents the stepwise regression analysis for screening out the three causal variables with the most significant functional contribution to the consequent variable soil stewardship perception. It has been found that the following variables has been retained at the last step and these arecommunication variable, total yield of crops cultivated, size of homestead land and family size. The r^2 value stands at 57.20 per cent which implies that these four variables collectively contributed 57.20per cent of variance in the consequent variable soil stewardship perception. The rest twenty variables have contributed only [(57.20/86.10) *100] per cent i.e., 66.43 per cent variance.

The fact has elicited from this hard evidence speaks that soil stewardship is precisely defined ecological, technological function. Exposure to mass media, and some training, getting through related literature, cross community communication etc. are so much important and all these are communication variable to build up competence in the domain of soil stewardship function in a farmer.

Stewardship functions when goes better, the yield will improve because the fertility status and management of the soil has got direct contribution to yield behavior of the crop.

Soil stewardship ignores the demarcation between main crop land or homestead land. In this area the vegetable cultivation is done in the homestead land or homestead surrounding land that is how it is related in terms of soil stewardship perception.

When the family size is higher, all the experiences are shared and that is how better possibility of all these stewardship function. Stewardship function is not only the individual farmer's function, it is basically a family function and today the family farming is a concept across the world, so no attributes can be isolated for other family members.

Table 5.33: Path Analysis: Decomposition of Total Effect into Direct, Indirect and Residual Effect: Soil stewardship perception(y_8)

Sl. No	Variables	Total Effect	Direct Effect	Indirect Effect	Highest Indirect Effect
1	Age (x_1)	0.195	-0.218	0.413	-0.570(x_{17})
2	Education (x_2)	0.238	0.236	0.002	0.104(x_{17})
3	Functional education (x_3)	-0.026	-0.437	0.411	0.144(x_{17})
4	Family size (x_4)	0.251	0.08	0.171	0.450(x_{17})
5	Size of land holding (bigha) (x_5)	0.148	-0.151	0.299	0.659(x_{17})
6	No. of land fragments (x_6)	0	0.062	-0.062	0.306(x_{17})
7	Average size of land fragment (katha) (x_7)	0.081	**-0.951**	**1.032**	0.715(x_{17})
8	Mean distance between two land fragments (m) (x_8)	-0.103	-0.154	0.051	0.716(x_{17})
9	Size of homestead land (katha) (x_9)	-0.008	0.395	-0.403	0.065(x_{17})
10	No. of crops cultivated (x_{10})	0.234	0.226	0.008	0.320(x_{17})
11	Total yield of crops cultivated (kg) (x_{11})	0.392	-0.451	0.843	-0.119(x_{17})
12	Total marketed surplus of crops cultivated (kg) (x_{12})	-0.183	0.355	-0.538	-0.519(x_{17})
13	Total input cost (Rs) (x_{13})	-0.098	-0.377	0.279	-0.293(x_{17})
14	On farm income (Rs) (x_{14})	-0.087	0.249	-0.336	0.677(x_{17})
15	Family expenditure/year (x_{15})	0.268	-0.223	0.491	0.132(x_{17})
16	Savings/year (x_{16})	-0.129	0.645	-0.774	0.706(x_{17})

Sl. No	Variables	Total Effect	Direct Effect	Indirect Effect	Highest Indirect Effect
17	Land under irrigation (bigha) (x_{17})	0.148	0.132	0.016	-0.100(x_5)
18	Cropping intensity (%) (x_{18})	0.319	-0.634	0.953	-0.798(x_{17})
19	Total hours of irrigation given (x_{19})	0.055	0.496	-0.441	0.212(x_{17})
20	Livestock count (x_{20})	0.008	-0.242	0.25	-0.866(x_{17})
21	Communication variable (x_{21})	-0.11	0.262	-0.372	-0.783(x_5)
22	Consumption of coal,firewood,fuelwood (kg/year) (x_{22})	-0.093	-0.592	0.499	-0.665(x_{17})
23	Production of organic manure (kg/year) (x_{23})	-0.149	-0.215	0.066	-0.492(x_{17})

Residual effect: 0.125

Result: Table 5.33 presents the path analysis wherein the total effect of exogenous variable on consequent variable has been decomposed into direct, indirect and residual effect.

Revelation: The causal variable average size of land has exerted both the highest direct and indirect effect on soil stewardship perception and the beta value indicates that with the increase in number and intensity of fragmentation it becomes difficult to manage the soil and uphold its fertility status towards attaining a sustainable character of agriculture. When fertility status is maintained, through soil stewardship crater by the farmer, it is expected that the ecological resilience of the agro-ecosystem as a whole will be maintained. It has also been observed that the farmers are becoming increasingly conscious about the deterioration of soil health, as reflected through low organic carbon content in the soil and they are increasingly conscious to follow soil stewardship through application of biofertilizers, testing of soil and increased adoption of mulching practices. No one is careful for a tiny and fragmented holding. This is the real difficulty being faced by the Indian farming. Soil health or betterment of soil health etc. are generally perceived as a function for the large size of holding. There is no such policy or even intentions among the farmers to feed their tiny and fragmented holding under all these stewardship functions.

The residual value is 0.125 it infers that a little over 12.5 per cent cannot be explain with this combination of 23 variables. It has been supported by the R square value 86.10 per cent as well.

It has been found that the exogenous variable land under irrigation has appeared as many as 21 out of 23 times in exerting the highest effect on the consequent variable soil stewardship perception, it implies that in an irrigated agro-based ecosystem the energy and input management system is very much sensitive and we have to be careful because number of irrigation involves energy. When energy is more, then the application of fertilizers will become more. Then again it has got energy equivalent and at the same time when it is intensive agriculture under irrigated agro-based ecosystem, the total number of labour engaged will be more, means it has got energy equivalent. So, that is how it has got tremendous associating effect or ability to influence characters of other variable.

Table 5.34: Coefficient of Correlation (r): Water stewardship perception (y_9)

Sl. No.	Independent Variables	'r' Value	Remarks
1	Age (x_1)	-0.171	
2	Education (x_2)	**0.511**	**
3	Functional education (x_3)	0.076	
4	Family size (x_4)	0.034	
5	Size of land holding (bigha) (x_5)	0.026	
6	No. of land fragments (x_6)	-0.225	
7	Average size of land fragment (katha) (x_7)	0.165	
8	Mean distance between two land fragments (m) (x_8)	0.068	
9	Size of homestead land (katha) (x_9)	0.185	
10	No. of crops cultivated (x_{10})	0.250	
11	Total yield of crops cultivated (kg) (x_{11})	**0.464**	**
12	Total marketed surplus of crops cultivated (kg) (x_{12})	0.168	
13	Total input cost (Rs) (x_{13})	0.094	
14	On farm income (Rs) (x_{14})	-0.019	
15	Family expenditure/year (x_{15})	**0.347**	*
16	Savings/year (x_{16})	0.004	
17	Land under irrigation (bigha) (x_{17})	-0.074	
18	Cropping intensity (%) (x_{18})	0.206	
19	Total hours of irrigation given (x_{19})	0.146	
20	Livestock count (x_{20})	0.123	
21	Communication variable (x_{21})	0.191	
22	Consumption of coal,firewood,fuelwood (kg/year) (x_{22})	0.014	
23	Production of organic manure (kg/year) (x_{23})	0.063	

**Correlation is significant at the 0.01 level

*Correlation is significant at the 0.05 level

Model 5.17: Coefficient of correlation: Water stewardship perception (y_9)and independent variables having significant correlation.

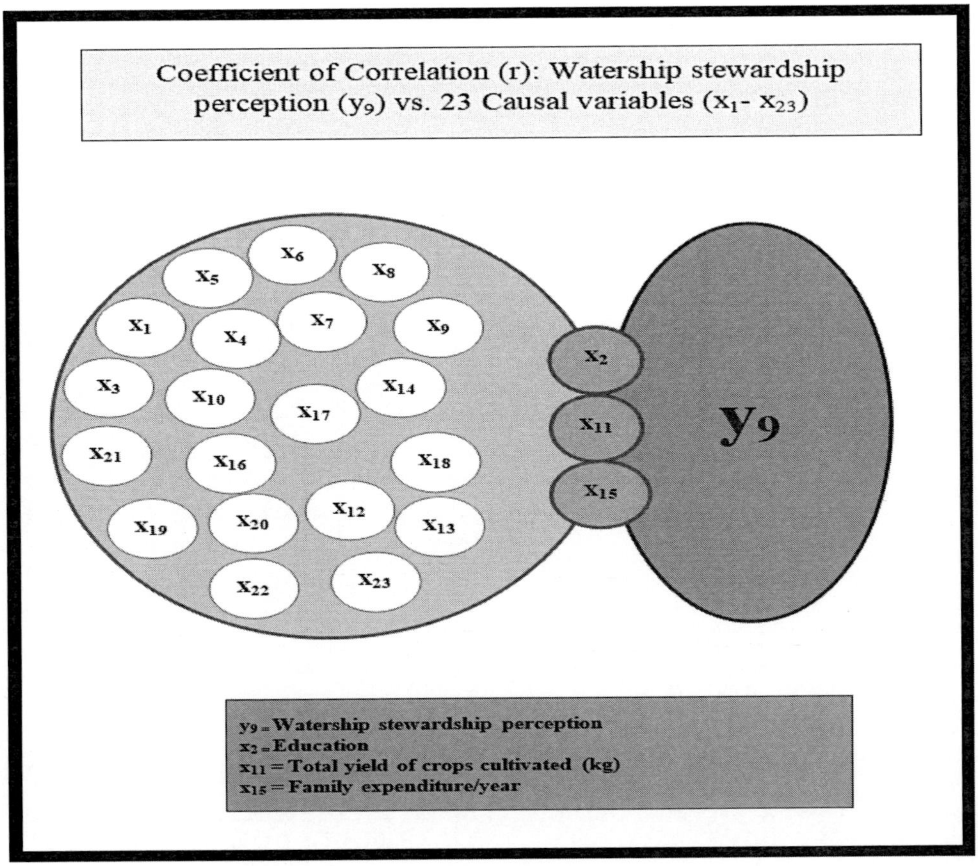

Result: Table 5.34 presents the coefficient of correlation between water stewardship perception (y_9) Vs 23 independent variables (x_1-x_{23}). It has been found that the following variable education (x_2) total yield of crops cultivated (kg) (x_{11}) and family expenditure/year (x_{15}) has recorded significant and positive correlation with the dependent variable, water stewardship perception (y_9).

Revelation: The variable education has recorded significant and positive correlation with water stewardship perception. This is quite obvious because water is an important determinant in an irrigated agro-ecosystem. It is impacted by the education of the family because water is gradually become a scarce commodity due to our wastage and indiscriminate use of groundwater. So a kind of general awareness has been build up in terms of water stewardship perception.

The variable total yield of crops cultivated has recorded significant and positive correlation with water stewardship perception. It implies that the yield of irrigated agro-ecosystem water is the prime determinant. Scientific management of water resources is the prime determinant in the yield behavior of the crop.

The variable family expenditure/year has recorded significant and positive correlation with water stewardship perception. It implies that the trajectory of the operating agro-ecosystem is heavily dependent on hydrodynamic in the form of water resources, management and its cost that is how cost or expenditure is related to the purchase of irrigation water and therefore, more is the wastage or unplanned consumption, the higher will be the expenditure.

Table 5.35: Multiple Regression Analysis: Water stewardship perception (y_9) Vs. 23 Causal Variables

Sl. No.	Variables	Reg. Coef. B	S.E. B	Beta	t Value
1	Age (x_1)	-0.035	0.018	**-0.658**	-1.989
2	Education (x_2)	0.081	0.095	0.260	0.857
3	Functional education (x_3)	-0.070	0.143	-0.218	-0.491
4	Family size (x_4)	-0.060	0.103	-0.157	-0.583
5	Size of land holding (bigha) (x_5)	0.013	0.076	0.046	0.174
6	No. of land fragments (x_6)	-0.001	0.008	-0.039	-0.095
7	Average size of land fragment (katha) (x_7)	-0.001	0.001	-0.261	-0.493
8	Mean distance between two land fragments (m) (x_8)	0.036	0.026	**0.544**	1.355
9	Size of homestead land (katha) (x_9)	0.066	0.139	0.191	0.473
10	No. of crops cultivated (x_{10})	4.710E-05	0.000	**0.749**	1.834
11	Total yield of crops cultivated (kg) (x_{11})	-7.583E-06	0.000	-0.184	-0.600
12	Total marketed surplus of crops cultivated (kg) (x_{12})	-1.142E-06	0.000	-0.133	-0.550
13	Total input cost (Rs) (x_{13})	-7.950E-07	0.000	-0.387	-1.111
14	On farm income (Rs) (x_{14})	-2.681E-07	0.000	-0.021	-0.070
15	Family expenditure/year (x_{15})	4.070E-07	0.000	0.031	0.120
16	Savings/year (x_{16})	-0.010	0.075	-0.065	-0.131
17	Land under irrigation (bigha) (x_{17})	-0.001	0.003	-0.160	-0.445
18	Cropping intensity (%) (x_{18})	-0.001	0.001	-0.278	-0.867
19	Total hours of irrigation given (x_{19})	0.080	0.079	0.312	1.022
20	Livestock count (x_{20})	-0.003	0.006	-0.195	-0.559
21	Communication variable (x_{21})	0.000	0.000	0.250	1.130
22	Consumption of coal,firewood,fuelwood (kg/year) (x_{22})	-3.355E-05	0.000	-0.121	-0.302
23	Production of organic manure (kg/year) (x_{23})	-0.035	0.018	**-0.658**	-1.989

R square: 85.20%

The standard error of the estimate: 0.471

Result: Table-5.35 presents the full model of multiple regression analysis between 23 causal variables (x_1-x_{23}) and water stewardship perception (y_9).

Revelation: The table-5.35 presents multiple regression analysis (full model) between 23 causal variables (x_1-x_{23}) and water stewardship perception (y_9). The three substantives impacts (Beta value) on consequent variable water stewardship perception are in order.

Highlights of the result

- Beta coefficient of the causal variable, no. of crops cultivated, is positive and significant on the consequent variable, water stewardship perception, it implies that water budget depends on the number of crops and types of the crops. When number of crops are more quite logically the consumption of water will be more and again it depends on the type of crop. When summer rice consumes huge and huge water, pulse and oilseed consumes less and less water.
- Both the causal variable, age and production of organic manure, have recorded same beta coefficient value and both the values are negative and significant. Chronological age means level of maturity and experience, the higher is the age, better has been the perception of water stewardship. Production of organic matter and water management they are highly integrated and correlated that is how both are impacting with equal frequency.
- Beta coefficient of the causal variable, mean distance between two land fragments is positive and significant. It implies that when the mean distance between two land fragments is higher, the water stewardship will be difficult.

The r^2 value being a combination of all these 23 variables has contributed 85.20 per cent of variance embedded with the consequent variable soil stewardship perception.

Table 5.36: Stepwise Regression Analysis: Water stewardship perception (y_9) Vs. 23 Causal Variables

Sl. No	Variables	Reg.coef. B	S.E. B	Beta	t value
1	Total yield of crops cultivated (kg) (x_{11})	4.974E-05	0.000	0.791	6.846

R square: 62.60%

The standard error of the estimate: 0.374

Model 5.18: Stepwise Regression Analysis: Water stewardship perception (y_9) Vs. 23 Causal Variables

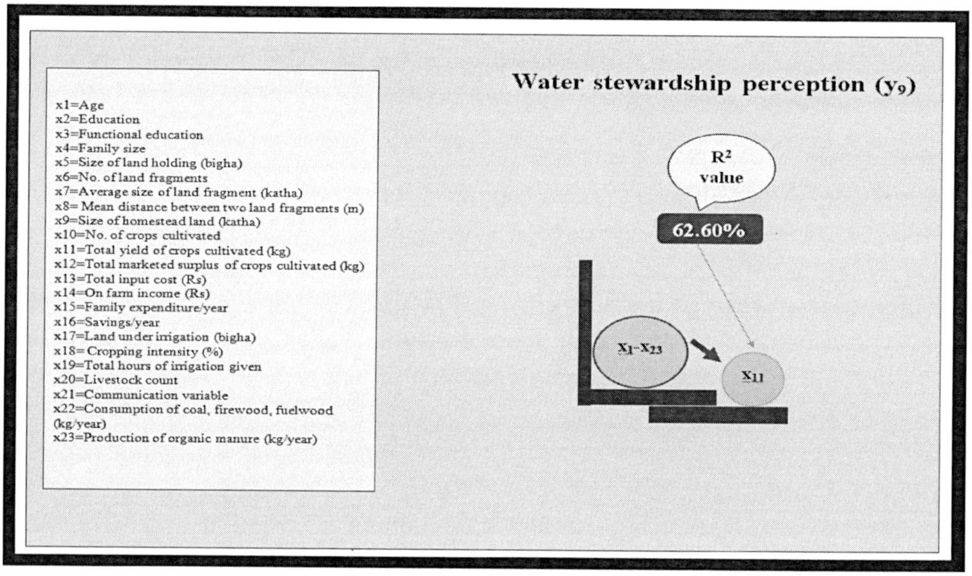

Result: Table 5.36 presents the stepwise regression analysis which elicits that one causal variable total yield of crops cultivated (kg) (x_{11}), has come out with stronger determining character on the consequent variable, water stewardship perception (y_9).

Revelation: Table 5.35 presents the stepwise regression analysis for screening out the one causal variable with the most significant functional contribution to the consequent variable water stewardship perception. It has been found that the following variable has been retained at the last step and that is total yield of crops cultivated. The r^2 value stands at 62.60 per cent which implies that the variable total yield of crops cultivated have contributed 62.60 per cent of variance in the consequent variable water stewardship perception. The rest twenty variables have contributed only [(62.60/85.20) *100] per cent i.e., 73.47per cent variance.

The fact has elicited from this hard evidence speaks that water stewardship functions when goes better, the yield will improve because better management of the water has got direct contribution to yield behavior of the crop.

Table 5.37: Path Analysis: Decomposition of Total Effect into Direct, Indirect and Residual Effect: Water stewardship perception(y9)

Sl. No	Variables	Total Effect	Direct Effect	Indirect Effect	Highest Indirect Effect
1	Age (x_1)	-0.171	-0.658	0.487	0.810(x_5)
2	Education (x_2)	0.511	0.260	0.251	-0.750(x_5)
3	Functional education (x_3)	0.076	-0.218	0.294	-0.388(x_5)

Sl. No	Variables	Total Effect	Direct Effect	Indirect Effect	Highest Indirect Effect
4	Family size (x_4)	0.034	-0.157	0.191	-0.136(x_5)
5	Size of land holding (bigha) (x_5)	0.026	0.045	-0.019	0.713(x_{17})
6	No. of land fragments (x_6)	-0.225	-0.039	-0.186	-0.826(x_5)
7	Average size of land fragment (katha) (x_7)	0.165	-0.261	0.426	-0.402(x_5)
8	Mean distance between two land fragments (m) (x_8)	0.068	0.544	-0.476	-0.522(x_5)
9	Size of homestead land (katha) (x_9)	0.185	0.191	-0.006	-0.395(x_5)
10	No. of crops cultivated (x_{10})	0.250	**0.747**	-0.497	-0.467(x_5)
11	Total yield of crops cultivated (kg) (x_{11})	0.464	-0.184	0.648	0.554(x_5)
12	Total marketed surplus of crops cultivated (kg) (x_{12})	0.168	-0.133	0.301	0.137(x_5)
13	Total input cost (Rs) (x_{13})	0.094	-0.387	0.481	-0.426(x_{17})
14	On farm income (Rs) (x_{14})	-0.019	-0.021	0.002	-0.128(x_5)
15	Family expenditure/year (x_{15})	0.347	0.031	0.316	-0.990(x_5)
16	Savings/year (x_{16})	0.004	-0.063	0.067	-0.410(x_5)
17	Land under irrigation (bigha) (x_{17})	-0.074	-0.160	0.086	-0.919(x_5)
18	Cropping intensity (%) (x_{18})	0.206	-0.278	0.484	0.804(x_5)
19	Total hours of irrigation given (x_{19})	0.146	0.312	-0.166	-0.227(x_5)
20	Livestock count (x_{20})	0.123	-0.192	0.315	0.763(x_5)
21	Communication variable (x_{21})	0.191	0.250	-0.059	-0.168(x_5)
22	Consumption of coal,firewood,fuelwood (kg/year) (x_{22})	0.014	-0.121	0.135	0.691(x_5)
23	Production of organic manure (kg/year) (x_{23})	0.063	-0.653	**0.716**	0.056(x_5)

Residual effect: 0.134

Result: Table 5.37 presents the path analysis wherein the total effect of exogenous variable on consequent variable has been decomposed into direct, indirect and residual effect.

Revelation: The causal variable, no. of crops cultivated, has exerted the highest direct effect on water stewardship perception and the beta value indicates that the water consuptions will increase with the increase in number of crops cultivated. The water is monitored, collected and disbursed through an easy access to groundwater, here in the zone, that is how both the yield and cropping intensity has been higher for this research locale. The crop diversity has also been impacted by the easy access to groundwater. The higher is the cropping intensity, the more is the count of crops per unit area and hence the water requirement and stewardship perception has gone up for the respondent.

The causal variable production of organic manure has exerted the highest indirect effect on water stewardship perception. It justifies that only water cannot solve the problem or restore ecological resilience unless it is associated with incorporation of crop residues and organic manure. Only water cannot supplement the nutrient requirement of the crops, water has got only a catalytic function.

The residual value is 0.134 it infers that a little over 13.4 per cent cannot be explain with this combination of 23 variables. It has been supported by the R square value 85.20 per cent as well.

It has been found that the exogenous variable size of land holding has appeared as many as 21 out of 23 times in exerting the highest effect on the consequent variable water conservation perception. The water management and use efficiency depends on the size of the land holding. When the size of land holding is large, it is quite likely that water use efficiency will be negatively affected. More water will go after waste and cost of irrigation will go up as well. The higher is the size of land holding, the higher would be the water consumption. The monitoring and management of water consumption here are being dictated by the size of land resources at large. The impact of size of land holding on water stewardship perception also impacted by the number of fragmentation held by the given size of land holding.

Table 5.38: Coefficient of Correlation (r): Biodiversity stewardship perception(y_{10})

Sl. No.	Independent Variables	'r' Value	Remarks
1	Age (x_1)	-0.028	
2	Education (x_2)	0.029	
3	Functional education (x_3)	**-0.303**	*
4	Family size (x_4)	0.117	
5	Size of land holding (bigha) (x_5)	-0.003	
6	No. of land fragments (x_6)	-0.149	
7	Average size of land fragment (katha) (x_7)	-0.092	
8	Mean distance between two land fragments (m) (x_8)	0.206	
9	Size of homestead land (katha) (x_9)	0.042	
10	No. of crops cultivated (x_{10})	0.090	
11	Total yield of crops cultivated (kg) (x_{11})	0.037	
12	Total marketed surplus of crops cultivated (kg) (x_{12})	-0.047	
13	Total input cost (Rs) (x_{13})	0.070	
14	On farm income (Rs) (x_{14})	0.188	
15	Family expenditure/year (x_{15})	0.097	

Sl. No.	Independent Variables	'r' Value	Remarks
16	Savings/year (x_{16})	**-0.418**	**
17	Land under irrigation (bigha) (x_{17})	-0.022	
18	Cropping intensity (%) (x_{18})	0.055	
19	Total hours of irrigation given (x_{19})	0.096	
20	Livestock count (x_{20})	**0.311**	*
21	Communication variable (x_{21})	0.059	
22	Consumption of coal,firewood,fuelwood (kg/year) (x_{22})	0.022	
23	Production of organic manure (kg/year) (x_{23})	-0.102	

**Correlation is significant at the 0.01 level

*Correlation is significant at the 0.05 level

Model 5.19: Coefficient of correlation: Biodiversity stewardship perception (y_{10}) and independent variables having significant correlation.

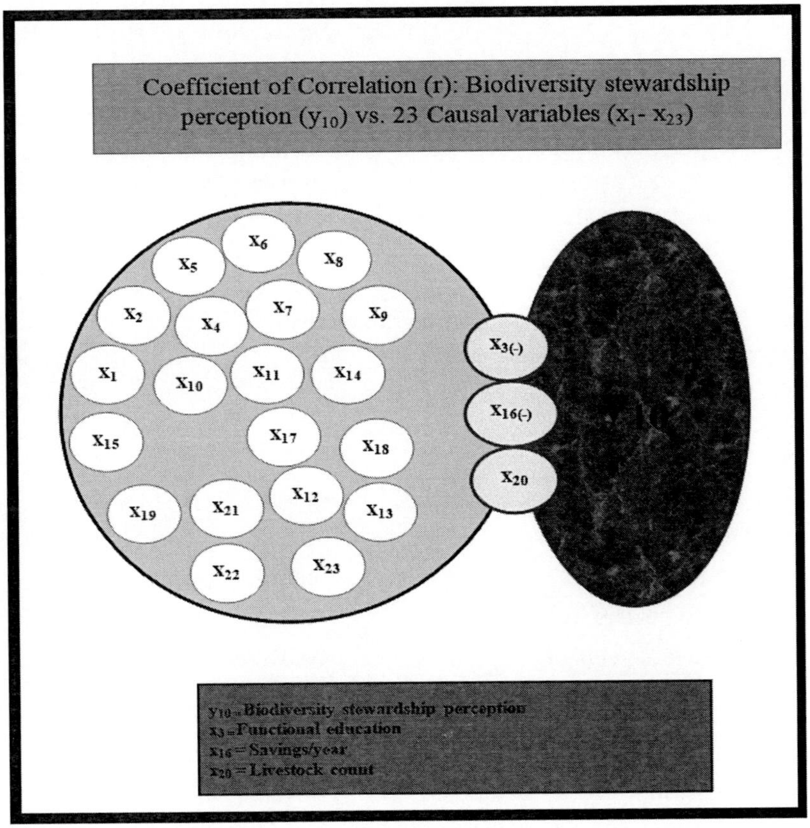

Result: Table 5.38 presents the coefficient of correlation between biodiversity stewardship perception (y_{10}) Vs 23 independent variables (x_1-x_{23}). It has been found that the following variable functional education (x_3), savings/year (x_{16})

has recorded significant but negative correlation with the dependent variable, biodiversity stewardship perception and the variable livestock count (x_{20}) has recorded significant and positive correlation with the dependent variable, biodiversity stewardship perception (y_{10}).

Revelation: The variable functional education has recorded significant and negative correlation with biodiversity stewardship perception. It implies that better functional education means better ecological learning and more concern about the biodiversity that is why training on traditional cultivar is attaining on higher ground and community gene bank have been created to mentor and nurture traditional crop variety. But in this research locale it has been found that most of the training are not purported to the ecological aspect that is why in such cases the functional education mostly hovering around the technological and input aspect so in that case we found discernible impact.

The variable savings/year has recorded significant but negative correlation with the dependent variable biodiversity stewardship perception. It implies that when biodiversity is less, yield will be inconsistent and to be followed by decreasing return to contribute in savings.

The variable livestock count has recorded significant and positive correlation with the dependent variable, biodiversity stewardship perception. This is quite obvious as biodiversity does not mean that it is only crop biodiversity. Biodiversity at large includes crops, animals, fish and microorganisms. When biodiversity is high means crop yield will be better and it will be sustainable.

Table 5.39: Multiple Regression Analysis: Biodiversity stewardship perception (y_{10}) Vs. 23 Causal Variables

Sl. No.	Variables	Reg. Coef. B	S.E. B	Beta	t Value
1	Age (x_1)	-0.012	0.020	-0.210	-0.565
2	Education (x_2)	0.064	0.111	0.198	0.581
3	Functional education (x_3)	-0.426	0.168	**-1.268**	-2.543
4	Family size (x_4)	-0.012	0.120	-0.029	-0.098
5	Size of land holding (bigha) (x_5)	-0.029	0.089	-0.097	-0.328
6	No. of land fragments (x_6)	-0.003	0.010	-0.143	-0.305
7	Average size of land fragment (katha) (x_7)	-0.002	0.001	-1.045	-1.756
8	Mean distance between two land fragments (m) (x_8)	0.050	0.031	0.730	1.618
9	Size of homestead land (katha) (x_9)	-0.094	0.162	-0.262	-0.577
10	No. of crops cultivated (x_{10})	-2.209E-05	0.000	-0.337	-0.736
11	Total yield of crops cultivated (kg) (x_{11})	-1.703E-05	0.000	-0.397	-1.152

Sl. No.	Variables	Reg. Coef. B	S.E. B	Beta	t Value
12	Total marketed surplus of crops cultivated (kg) (x_{12})	-1.328E-06	0.000	-0.148	-0.546
13	Total input cost (Rs) (x_{13})	-3.513E-07	0.000	-0.164	-0.420
14	On farm income (Rs) (x_{14})	2.839E-06	0.000	0.209	0.629
15	Family expenditure/year (x_{15})	-1.082E-05	0.000	-0.801	-2.720
16	Savings/year (x_{16})	0.166	0.087	**1.059**	1.895
17	Land under irrigation (bigha) (x_{17})	0.008	0.003	1.027	2.538
18	Cropping intensity (%) (x_{18})	-0.001	0.001	-0.479	-1.332
19	Total hours of irrigation given (x_{19})	0.065	0.092	0.241	0.703
20	Livestock count (x_{20})	0.014	0.007	0.774	1.975
21	Communication variable (x_{21})	0.000	0.000	0.383	1.539
22	Consumption of coal,firewood,fuelwood (kg/year) (x_{22})	0.000	0.000	**-1.067**	-2.378
23	Production of organic manure (kg/year) (x_{23})	-0.012	0.020	-0.210	-0.565

R square: 81.30%

The standard error of the estimate: 0.551

Result: Table-5.35 presents the full model of multiple regression analysis between 23 causal variables (x_1-x_{23}) and biodiversity stewardship perception (y_{10}).

Revelation: The table-5.35 presents multiple regression analysis (full model) between 23 causal variables (x_1-x_{23}) and biodiversity stewardship perception (y_{10}). The three substantives impact (Beta value) on consequent variable biodiversity stewardship perception are in order.

Highlights of the result

- Beta coefficient of the causal variable, functional education, is negative and significant on the consequent variable biodiversity stewardship perception. It implies that better functional education means better ecological learning and more concern about the biodiversity, that is why training on traditional cultivar is attaining the higher ground and community gene banks have been created to mentor and nurture traditional crop variety. But in this research locale, it has been found that most of the training are not purported to the ecological aspect that is why in such cases the functional education is mostly hovering around the technological and input aspect and are found discernible impact.
- Beta coefficient of the causal variable consumption of coal, firewood, fuelwood is negative and significant on the consequent variable, biodiversity stewardship perception, it implies that energy consumption and biodiversity

stewardship perception has got an inverse relation. The more energy we consume, the less will be the biodiversity because energy sources in most cases especially in rural areas, the bio-energy sources are different trees, plants etc. and rural people are consuming firewood, fuelwood and at the same time they are creating stress in biodiversity.

- Beta coefficient of the causal variable savings/year is positive and significant on the consequent variable biodiversity stewardship perception, it implies that naturally when biodiversity is more, yield will be consistent and to be followed by a persistent increasing return to contribute in savings.

The r^2 value being a combination of all these 23 variables has contributed 81.30 per cent of variance embedded with the consequent variable biodiversity stewardship perception.

Table 5.40: Stepwise Regression Analysis: Biodiversity stewardship perception (y_{10}) Vs. 23 Causal Variables

Sl. No	Variables	Reg.coef. B	S.E. B	Beta	t value
1	Savings/year (x_{16})	-6.681E-06	0.000	-0.494	-3.139
2	Livestock count (x_{20})	0.101	0.042	0.377	2.394

R square: 34.00%

The standard error of the estimate: 0.527

Model-6.20: Stepwise Regression Analysis: Biodiversity stewardship perception (y_{10}) Vs. 23 Causal Variables

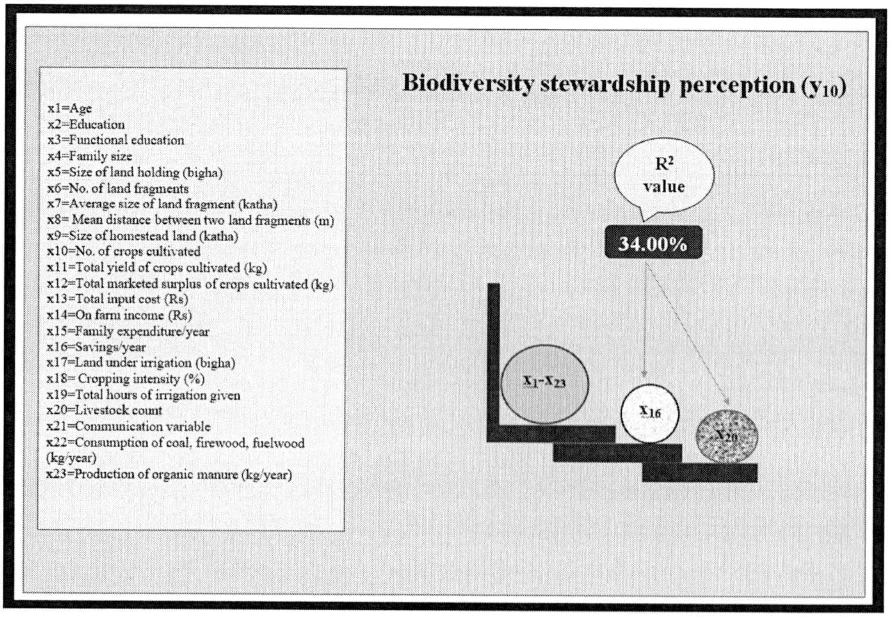

Result: Table 5.39 presents the stepwise regression analysis which elicits that two causal variable savings/year (x_{16}) and livestock count (x_{20}) has come out with stronger determining character on the consequent variable, biodiversity stewardship perception (y_{10}).

Revelation: Table 5.39 presents the stepwise regression analysis for screening out of the two causal variables with the most significant functional contribution, to the consequent variable, biodiversity stewardship perception. It has been found that the following variable has been retained at the last step and that is total yield of crops cultivated. The r^2 value stands at 34.00 per cent, which implies that the variable savings/year and livestock count collectively contributed 34.00 per cent of variance in the consequent variable, biodiversity stewardship perception. The rest twenty variables have contributed only [(34.00/81.30) *100] per cent i.e., 41.82per cent variance.

The fact has elicited from this hard evidence speaks that when biodiversity is less, yield will be inconsistent, and to be followed by decreasing return to contribute in savings.

Livestock count is so important in restoring ecological resilience basically management of operating farm. Livestock are source of organic carbon, if the count is declining then the organic matter will be reduced and source of organic carbon will be disrupted and that is the problem. Only crop cultivation cannot make a progress unless it is supported by livestock and fisheries.

Table 5.41: Path Analysis: Decomposition of Total Effect into Direct, Indirect and Residual Effect: Biodiversity stewardship perception (y_{10})

Sl. No	Variables	Total Effect	Direct Effect	Indirect Effect	Highest Indirect Effect
1	Age (x_1)	-0.028	-0.210	0.182	-0.104(x_5)
2	Education (x_2)	0.029	0.198	-0.169	0.013(x_5)
3	Functional education (x_3)	-0.303	**-1.268**	0.965	0.180(x_5)
4	Family size (x_4)	0.117	-0.029	0.146	0.282(x_5)
5	Size of land holding (bigha) (x_5)	-0.003	-0.096	0.093	-0.619(x_{17})
6	No. of land fragments (x_6)	-0.149	-0.14	-0.009	0.949(x_5)
7	Average size of land fragment (katha) (x_7)	-0.092	-1.045	0.953	0.910(x_5)
8	Mean distance between two land fragments (m) (x_8)	0.206	0.73	-0.524	-0.451(x_5)
9	Size of homestead land (katha) (x_9)	0.042	-0.261	0.303	0.518(x_5)
10	No. of crops cultivated (x_{10})	0.09	-0.337	0.427	0.535(x_{18})
11	Total yield of crops cultivated (kg) (x_{11})	0.037	-0.397	0.434	-0.782(x_5)

Sl. No	Variables	Total Effect	Direct Effect	Indirect Effect	Highest Indirect Effect
12	Total marketed surplus of crops cultivated (kg) (x_{12})	-0.047	-0.148	0.101	-0.139(x_5)
13	Total input cost (Rs) (x_{13})	0.070	-0.164	0.234	0.512(x_9)
14	On farm income (Rs) (x_{14})	0.188	0.209	-0.021	0.332(x_5)
15	Family expenditure/year (x_{15})	0.097	-0.801	0.898	0.493(x_5)
16	Savings/year (x_{16})	-0.418	1.059	**-1.477**	0.805(x_5)
17	Land under irrigation (bigha) (x_{17})	-0.022	1.027	-1.049	0.780(x_5)
18	Cropping intensity (%) (x_{18})	0.055	-0.477	0.532	-0.824(x_5)
19	Total hours of irrigation given (x_{19})	0.096	0.241	-0.145	0.805(x_5)
20	Livestock count (x_{20})	0.311	0.774	-0.463	-1.688(x_5)
21	Communication variable (x_{21})	0.059	0.383	-0.324	0.713(x_5)
22	Consumption of coal,firewood,fuelwood (kg/year) (x_{22})	0.022	-1.066	1.088	-0.921(x_5)
23	Production of organic manure (kg/year) (x_{23})	-0.102	-0.209	0.107	-0.093(x_5)

Residual effect: 0.182

Result: Table 5.40 presents the path analysis wherein the total effect of exogenous variable on consequent variable has been decomposed into direct, indirect and residual effect.

Revelation: The causal variable functional education has exerted the highest direct effect on biodiversity stewardship perception and the beta value indicates that conventional education has got no value or impact rather capacity building in the form of training and exposure visits will help to create a much more impact on the perception of biodiversity stewardship of the respondents.

The causal variable savings/year has exerted the highest indirect effect on biodiversity stewardship perception and the beta value indicates that all these ecological jargoning and applications of ecological sermons will be executed only when the farmers will have ample of money at their disposal. If his/her farming occupation cannot provide ample income, the farmers nowhere will be interested for following ecological practices or its execution.

The residual value is 0.182 it infers that a little over 18.2 per cent cannot be explain with this combination of 23 variables. It has been supported by the R square value 81.30 per cent as well.

It has been found that the exogenous variable size of land holding has appeared as many as 20 out of 23 times in exerting the highest effect on the consequent variable biodiversity stewardship perception. This is quite obvious because the more is the size of land holding, the higher is the favourable effect on biodiversity.

Small and fragmented land holding does not allow diversity, floral and faunal to grow up and evolve. It is due to tiny size of operating land, there is no scope to have a pond or fish of marshy land along side of the main cropping unit. So the diversity has been negatively impacted with the more number of fragmentations.

FACTOR ANALYSIS

Table 5.42: Factor Analysis: Strategic Conglomeration of Variables into Factor

Factors	Variables	Factor loading	% of variance	Cumulative %	Factor renamed
1	Size of land holding (bigha) Average size of land fragment (katha) Mean distance between two land fragments (m) Land under irrigation (bigha)	0.910 0.677 0.646 0.909	16.505	16.505	***Farm ecology***
2	Total yield of crops cultivated (kg) Livestock count Communication variable Application of organic manure (kg/bigha/year)	-0.667 0.534 0.618 0.594	11.911	28.416	***Farm resource composition***
3	No. of crops cultivated Soil conservation perception	0.774 0.572	9.636	38.052	***Soil ecology***
4	On farm income (Rs) Family expenditure/year	0.521 -0.514	8.280	46.332	***Economic potency***
5	Size of homestead land (katha) Rice stubble height left in the field (inches) Biodiversity stewardship perception	0.548 -0.560 -0.550	7.569	53.901	***Conservation ecosystem***
6	Climate resilient technology Biodiversity conservation perception Soil stewardship perception Water stewardship perception	-0.593 -0.630 -0.624 -0.552	7.063	60.964	***Ecological resilience***
7	Total input cost (Rs)	0.529	5.297	66.261	
8	Crop residue leftin field (kg)	0.636	5.109	71.370	

Factors	Variables	Factor loading	% of variance	Cumulative %	Factor renamed
9	Total hours of irrigation given	0.567	4.453	75.823	
10	Age Functional education	-0.661 0.668	3.617	79.440	***Experiential learning***

Result: Table- 5.41 presents the strategic conglomeration of variables into factor through factor analysis.

Revelation: Factor analysis (Principle Component Analysis) is a data reduction approach. Here the total number of interacting variable, based on factor loading and Eigen values, has been grouped into identifiable factor. Each factor is retaining the homogeneous variable, strongly bound to each other, towards taking a strategic decision for effective management of resource.

Factor 1 has accommodated the following variables,**size of land holding (bigha) average size of land fragment (katha), mean distance between two land fragments (m), land under irrigation (bigha)** and has been renamed as ***farm ecology***. These factors have contributed 16.505 percent of variance in the total interaction.

Factor 2 has accommodated the following variables, **total yield of crops cultivated (kg), livestock count, communication variable, application of organic manure (kg/bigha/year)**and has been renamed as ***farm resource composition.*** These factors have contributed 11.911 percent of variance in the total interaction.

Factor 3 has accommodated the following variables, **no. of crops cultivated, soil conservation perception** and has been renamed as ***soil ecology***. These factors have contributed 9.636 percent of variance in the total interaction.

Factor 4 has accommodated the following variables, **on farm income (Rs) family expenditure/year** and has been renamed as ***economic potency.*** These factors have contributed 8.280 percent of variance in the total interaction.

Factor 5 has accommodated the following variables, **size of homestead land (katha), rice stubble height left in the field (inches), biodiversity stewardship perception** and has been renamed as ***conservation ecosystem.*** These factors have contributed 7.569 percent of variance in the total interaction.

Factor 6 has accommodated the following variables, **climate resilient technology, biodiversity conservation perception, soil stewardship perception, water stewardship perception** and has been renamed as ***ecological resilience.*** These factors have contributed 7.063 percent of variance in the total interaction.

Since **Factor 7,8 and 9** has retained only a solitary variable, there is no need for renaming so, it is left as such.

Factor 10 has accommodated the following variables, **age, functional education** and has been renamed as ***experiential learning.*** These factors have contributed 3.617 percent of variance in the total interaction.

Artificial Neural Network

Model 5.21: Artificial neural network

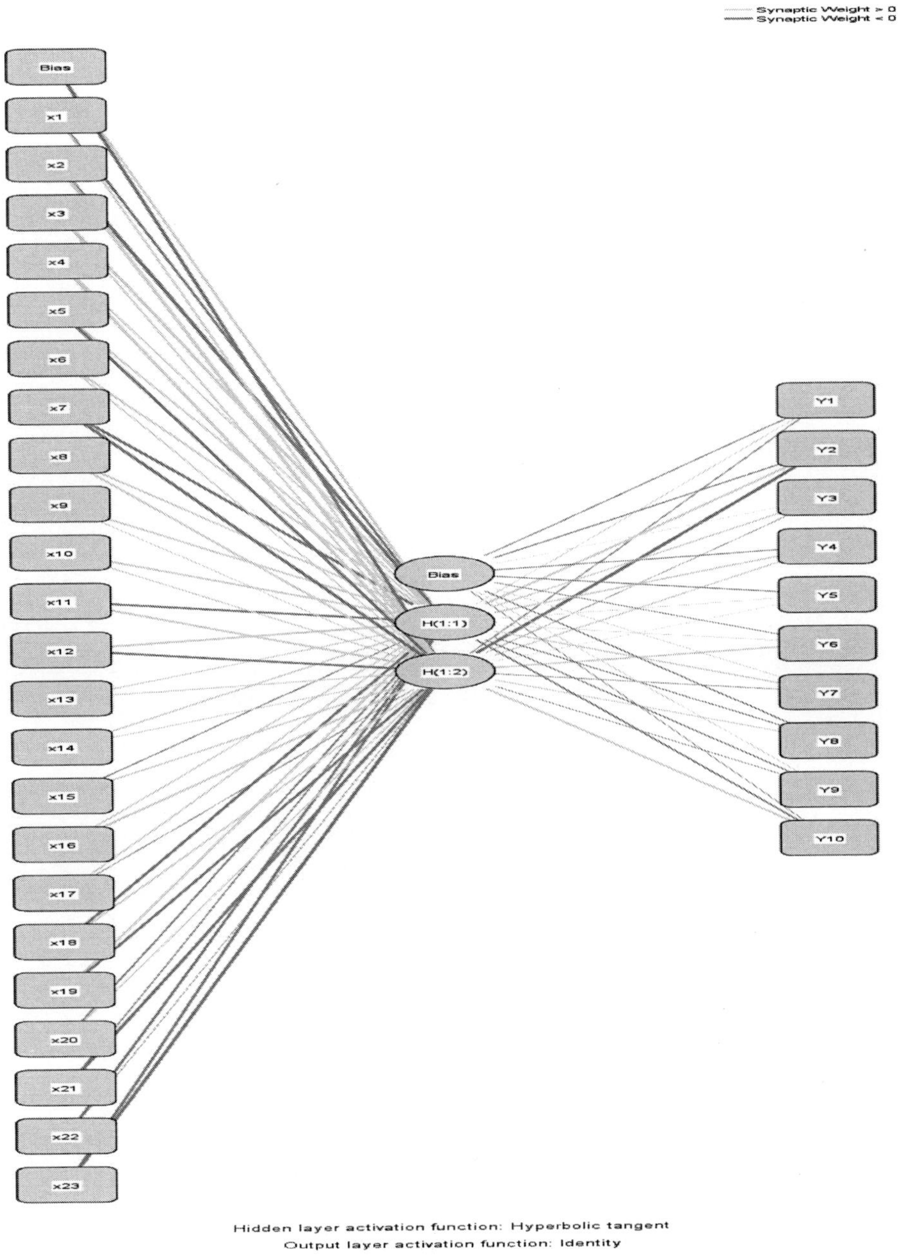

Table 5.43: Model summary of artificial neural network

Model Summary			
Training	Sum of Squares Error		103.810
	Average Overall Relative Error		0.944
	Relative Error for Scale Dependents	Application of organic manure (kg/bigha/year)	1.009
		Rice stubble height left in the field (inches)	0.709
		Crop residue left in field (kg)	0.960
		Climate resilient technology	0.974
		Soil conservation perception	1.001
		Water conservation perception	0.881
		Biodiversity conservation perception	0.959
		Soil stewardship perception	0.941
		Water stewardship perception	1.010
		Biodiversity stewardship perception	0.994
	Stopping Rule Used		1 consecutive step(s) with no decrease in error
	Training Time		0:00:00.01
Testing	Sum of Squares Error		24.343
	Average Overall Relative Error		1.041
	Relative Error for Scale Dependents	Application of organic manure (kg/bigha/year)	6.451
		Rice stubble height left in the field (inches)	0.691
		Crop residue left in field (kg)	1.028
		Climate resilient technology	0.934
		Soil conservation perception	1.023
		Water conservation perception	0.925
		Biodiversity conservation perception	1.168
		Soil stewardship perception	1.150
		Water stewardship perception	2.609
		Biodiversity stewardship perception	0.976
Error computations are based on the testing sample.			

Result: Model 5.19 depicts the artificial neural network analysis of 10 output variables (y_1-y_2)vs. 23 input variables.

Revelation: Artificial neural network is a mimicry of human brain. It is a part of artificial intelligence. Here 3 layers are there, first one is labelled as input layer, middle one is labelled as hidden layer and the last one is labelled as output layer.

The input layer of a neural network is composed of artificial input neurons, andbrings the initial data into the system for further processing by subsequent layers of artificial neurons. The input layer is the very beginning of the workflow for the artificial neural network.

In neural networks, a hidden layer is located between the input and output of the algorithm, in which the function applies weights to the inputs and directs them through an activation function as the output. Hidden layer are there to reduce the error by inducing an artificial value. In short, the hidden layersperform nonlinear transformations of the inputs entered into the network.

The output layer in an artificial neural network is the last layer of neurons that produces given outputs for the program. Though they are made much like other artificial neurons in the neural network, output layer neurons may be built or observed in a different way, given that they are the last "actor" nodes on the network.

The following input variables age (x_1), education (x_2), average size of land fragment (katha) (x_7), total yield of crops cultivated (kg) (x_{11}), cropping intensity (%) (x_{18}), livestock count (x_{20}), consumption of coal, firewood, fuelwood (kg/year) (x_{22}), production of organic manure (kg/year) (x_{23})after passing through the hidden layer 1 and has directly impacted on output variable soil stewardship perception (y_8), water stewardship perception (y_9) and biodiversity stewardship perception (y_{10}).

The following input variables size of land holding (bigha) (x_5), average size of land fragment (katha) (x_7), total marketed surplus of crops cultivated (kg) (x_{12}), total hours of irrigation given (x_{19}), livestock count (x_{20}), communication variable (x_{21}),consumption of coal, firewood, fuelwood (kg/year) (x_{22}), production of organic manure (kg/year) (x_{23}) after passing through the hidden layer 2 and has directly impacted on output variable application of organic manure (kg/bigha/year) (y_1), rice stubble height left in the field (inches) (y_2), crop residue left in field (kg) (y_3), biodiversity conservation perception (y_7) and water stewardship perception (y_9).

The highest impact has been on output variable rice stubble height left in the field (inches) (y_2).

The variables average size of land fragment (katha) (x_7), livestock count (x_{20}), consumption of coal, firewood, fuelwood (kg/year) (x_{22}), production of organic manure (kg/year) (x_{23})are the strongest determinant in characterizing these output variables application of organic manure (kg/bigha/year) (y_1), rice stubble height left in the field (inches) (y_2), crop residue left in field (kg) (y_3), biodiversity conservation perception (y_7), soil stewardship perception (y_8), water stewardship perception (y_9) and biodiversity stewardship perception (y_{10}).

It is clear that from the training variable the error sum of square has been substantially downsize in testing variable while passing through the hidden layer.

6

Summary and Conclusion

Without the earth's biological diversity, our lives would, literally, be nothing. In both the natural and built environment, biological diversity underpins ecosystems, which in turn support all the activities on which we depend for health, well-being, prosperity and inspiration. Healthy, stable and biodiverse ecosystems are the foundation of our health and well-being, as well as that of our fellow species. They help to regulate our climate and control extreme events, pests and diseases, as well as to provide us with water, food, raw materials and spaces for recreation. They absorb our wastes, sustain economic sectors and the livelihoods of millions of people, and they nurture our health, culture and spiritual fulfilment.

The need to restore damaged ecosystems has never been greater. Degradation is undermining hard-won development gains and threatening the well-being of today's youth and future generations, while making national commitments increasingly more difficult and costly to reach. We need to re-create a balanced relationship with nature, not only by conserving ecosystems that are still healthy, but also by urgently and sustainably restoring degraded ones. Ecosystem restoration alone cannot solve the crises we face, but it is key to averting the worst of them.

CA is an agro-ecological approach to sustainable production intensification. It involves the application of three inter-linked principles that underpin agricultural production systems based on locally formulated practices: (i) permanent no or minimum mechanical soil disturbance, which in practice entails direct seeding through mulch into no-till soils; (ii) maintenance of soil cover with crop residues and green manure crops, particularly legumes; and (iii) diversified cropping system involving annuals and perennial in rotations, sequences and associations.

In 2011, CA had spread over 125 million hectares (9% of the global cropped land) across all continents and most agro-ecologies, including small and large farms. In addition, there is a significant area of CA orchards in the Mediterranean countries. CA is now considered to be a practical agro-ecological approach to achieving sustainable agriculture intensification. It offers environmental, economic and social advantages that are not fully possible with tillage-based production systems, as well as improved productivity and resilience, and improved ecosystem services while minimizing the excessive use of agrochemicals, energy and heavy machinery. While there are challenges to the adoption of CA, there is also increasing interest from producers, the civil society, donors and private sector institutions to further promote and service the uptake and spread of CA globally.

Ecosystem services are mainly, benefits to humans from the ecosystems. Ecosystem services were classified into the following categories of services by the Millennium Ecosystem Assessment (2005). Ecosystem services to which conservation agriculture may have significant contributions are-

Provisioning services: These are ecosystem products or functions that support different life systems in and around

- food (including seafood and game), crops, wild foods, and spices
- water
- pharmaceuticals, biochemicals, and industrial products
- energy (hydropower, biomass fuels)

Regulating services: These are ecosystem functions that make the system deterrent of impurities, pollutants and regulate inhibitors to ecosystem resilience.

- carbon sequestration and climate regulation
- soil moisture storage
- regulation of stream flows and groundwater levels
- waste decomposition and detoxification
- purification of water and air
- crop pollination pest and disease control
- erosion control

Supporting services: The services that make the ecology more conducive to system productivity and resilience.

- nutrient dispersal and cycling
- seed dispersal
- primary production

Cultural services: This is related to functioning that supports aesthetics, entertainment, values and cultural ethos of community people and stakeholders.

- cultural, intellectual and spiritual inspiration
- recreational experiences (including ecotourism)
- scientific discovery

These are the basic ecological functions which are supporting the production and function of the entire agro-ecosystem, that is how it was essential to know to what extent and how these perception and stewardship functions are imparting on-

- Soil conservation perception
- Water conservation perception
- Biodiversity conservation perception
- Soil stewardship perception
- Water stewardship perception
- Biodiversity stewardship perception

The study has also included other predicted characters and those are application of organic manure, rice stubble height left in the field, crop residue left in the field and climate resilient technology. From this study it has been reflected that the following variables have been retained in the stepwise regression analysis and these are-

- Application of organic manure (y_1)- Livestock count (x_{20}), Size of homestead land (x_9), Size of land holding (x_5), Production of organic manure (x_{23}), Education (x_2)
- Rice stubble height left in the field (y_2)- Total input cost (x_{13}), Age (x_1), Functional education (x_3)
- Crop residue left in the field (y_3)- Communication variable (x_{21}), Mean distance between two land fragments (x_8)
- Climate resilient technology (y_4)- Consumption of coal, firewood, fuelwood (x_{22}), Production of organic manure (x_{23}), Total marketed surplus of crops cultivated (x_{12}), Total yield of crops cultivated (x_{11})
- Soil conservation perception (y_5)- Total hours of irrigation given (x_{19}), On farm income (x_{14})
- Water conservation perception (y_6)- No. of crops cultivated (x_{10}), Production of organic manure (x_{23}), Average size of land fragment (x_7)
- Biodiversity conservation perception (y_7)- Average size of land fragment (x_7), Total yield of crops cultivated (x_{11})
- Soil stewardship perception (y_8)- Communication variable (x_{21}), Total yield of crops cultivated (x_{11}), Size of homestead land (x_9), Family size (x_4)
- Water stewardship perception (y_9)- Total yield of crops cultivated (x_{11})
- Biodiversity stewardship perception (y_{10})- Savings/year (x_{16}), Livestock count (x_{20})

So, these variables are treated as mother variable which has got tremendous impact in monitoring and auditing of the ecological functions and services of the operating agro-ecosystem.

The main problem of the agriculture for the research locale under study is that ecological services are being disrupted and impaired due to the indiscriminate depletion and decline of natural resources like soil, water and biodiversity. If these are depleted, then the agro-ecology will lose its resilience and productive functions gradually. Then all these things will remain and soil will look like soil but not a handful grain will be produced there from; that is the total disaster we are going to face, unless-

i. Managing soil, water and biodiversity from ecological perspective.

ii. We can countervail the adverse impact of fragmentation through inviting community management of agro-ecosystem for e.g. community seed bed, community water management, community nursery etc. that would help to protect the soil, water, biodiversity and all.

iii. We need to have a policy change both at the macro-sociological and micro-sociological perspective.

These variables may be relevant for this research locale but may not be that relevant for other research locale. So this important functional research need to be replicated atleast for each block taking some selected farms and farmers.

Ultimately, the research suggests that an ecological policy should be incorporated in the technology adaptation process to make our agriculture more resilient like creation of community gene banks, community water auditing system, soil mulching and protecting the soil from erosion, crop diversification so that there can be a natural source of nitrogen, phosphate and potash in the soil. This complete approach through the agro-ecological functioning will take agriculture and humanity together. The impacts of the COVID-19 pandemic will be felt for generations. Yet this crisis has also demonstrated the power of international cooperation and provided us with an opportunity to steer away from our current destructive trajectory. To put countries on a path that is green, sustainable and fair, national governments must include ecosystem restoration in their pandemic recovery plans. This Decade can serve as a Launchpad to accelerate the transformative changes we need to combat the climate crisis, prevent mass extinctions and build social and economic resilience.

Recommendations

- Stewardship functions and conservation education should be incorporated in the present course and curriculum in order to generate basic and advance knowledge.
- The farm innovation like community gene bank, people's biodiversity registrar, folk gene consortia, soil conservation task force have to be created with necessary legal as well as statutory provision.
- Both up scaling and backstopping of technologies are essential for making stewardship function a sustainable proposition and practices.
- Agricultural researches along with extension researches should go for frame working, resource monitoring and stewardship functions therewith.

- ☐ Baseline surveys and bench marking information are essentials as to what and how the farm energy, fertilizer and agri chemicals, water and biodiversity are exploited, regenerated or eroded to frame up a community level policy with a proper justification for stewardship functions.
- ☐ Auditing, costing and financing after stewardship function should also be matched or complied with ecological deliverables generated by respective stewardship functions at present and in posterity.
- ☐ High end statistical packages are to be applied for efficient and high end monitoring of erosion depletion or retrieval of natural resources.

The present study has well germinated the needs of future researches and these are:

- ☐ Application of artificial intelligence for future simulation.
- ☐ Replication in more geo-spatial and resource location could have been helpful for better understanding.
- ☐ Ecological resilience should be entangled with social adaptability and economic viability.
- ☐ The future research invites agro-ecological zone wise data base creation and scenario generation which can be a part of policy research.

Bibliography

Acharya, S.K. and Chatterjee, Riti (2021)*Dynamics of Conservation Agriculture: a societal perspective,* Biodiversity and Conservation @ Springer, Review Paper, Published online: 22 March 2021

Acharya, S.K. and Chatterjee, Riti (2019) *Conservation Agriculture: The dynamics of Ecology and Ecological Services,* Indian Journal of Agriculture Business Volume 5 Number 2, July- December 2019

Adger, W. Neil (2000) *Social and ecological resilience: are they related?* Progress in Human Geography 24,3 (2000) pp. 347–364, Publishing Year- September 1, 2000

Agar, John W. M.; Simmonds, Rosemary E.; Knight, Richard and Somerville, Christine A. (2009) *Using water wisely: New, affordable, and essential water conservation practices for facility and home hemodialysis,* Hemodialysis International 2009; 13:32–37, Published on-2009

Ahmed, Saleh and Cokinos,Christopher (2017) *How does ecological modernization explain agriculture adaptation in coastal Bangladesh? A critical discussion,*Environmental Hazards@ Taylor Francis,16:2,133-148, Published on- 19 January 2017

Andersson, Jens A.; D'Souza, Shereen (2013) *From adoption claims to understanding farmers and contexts: A literature review of Conservation Agriculture (CA) adoption among smallholder farmers in southern Africa,* Agriculture, Ecosystems and Environment 187 (2014) 116–13, Published online- 21 September 2013

Arfanuzzaman, Md. & Atiq Rahman,A. (2017*) Sustainable water demand management in the face of rapid urbanization and ground water depletion for social–ecological resilience building,* Elsevier Science, Published on- 2017

Awasthi, Amit; Singh, Nirankar; Mittal, Susheel; Gupta, Prabhat K.; Agarwal, Ravinder (2010) *Effects of agriculture crop residue burning on children and young on PFTs in North West India,* Science of the Total Environment, Volume 408, Issue 20,15 September 2010, Pages 4440-4445, @ ScienceDirect, Elsevier, Published online- 16 July 2010

Ballard, Heidi L. & Jill M. Belsky,(2010) *Participatory action research and environmental learning: implications for resilient forests and communities,* Environmental

Education Research, 16:5-6, 611-627, DOI: 10.1080/13504622.2010.505440, Published online: 20 Oct 2010

Belay, Sisay A.; Assefa, Tewodros T.; Prasad, P. V. Vara; Schmitter, Petra; Worqlul, Abeyou W.; Steenhuis, Tammo S.; Reyes, Manuel R. and Tilahun, Seifu A. (2020) *The Response of Water and Nutrient Dynamics and of Crop Yield to Conservation Agriculture in the Ethiopian Highlands,* Sustainability 2020, 12, 5989, Published on- 24 July 2020

Bharti, Chandrabhan; Mohapatra, Anita; Maurya, Rakesh; Maharana, Chandan; Maurya, Alok and Malakar, Pranab (2020) *Nitrate pollution with modernization of Indian agriculture,* Journal of Pharmacognosy and Phytochemistry 2020; 9(3): 2073-2080, E-ISSN: 2278-4136 P-ISSN: 2349-8234, Published on- 2020

Busari, Mutiu Abolanle; Kukal, Surinder Singh; Kaur, Amanpreet; Bhatt, Rajan and Dulazi, Ashura Ally (2015) *Conservation tillage impacts on soil, crop and the environment,* International Soil and Water Conservation Research (2015), http://dx.doi.org/10.1016/j.iswcr.2015.05.002, Published on- 2015

Chapin, F Stuart; Sommerkorn, Martin; Robards, Martin D.; Hillmer-Pegram, Kevin (2015) *Ecosystem stewardship: A resilience framework for arctic conservation,* Elsevier Science, Published on 2015

Chapin, F Stuart; Carpenter, Stephen R.; Kofinas, Gary P.; Folke, Carl; Abel, Nick; Clark, William C.; Olsson, Per; Smith, D. Mark Stafford; Walker, Brian; Young, Oran R.; Berkes, Fikret; Biggs, Reinette; Grove, J. Morgan; Naylor, Rosamond L.; Pinkerton Evelyn; Steffen, Will; Swanson, Frederick J. (2010)*Ecosystem stewardship: sustainability strategies for a rapidly changing planet,* CelPress, Trends in Ecology & Evolution, @ScienceDirect Elsevier, Volume 25, Issue 4,April 2010,Pages 241-249

Chamala, Shankariah (1990) Social and environmental impacts of modernization of agriculture in developing countries, Environmental Impact Assessment Review, Volume 10, Issues 1–2,March–June 1990, Pages 219-231, Publishing Year- 1990

Cross, Molly S.; McCarthy, Patrick D.; Garfin, Gregg; Gori, David and Enquist, Carolyn A. F. (2012)*Accelerating Adaptation of Natural Resource Management to Address Climate Change,* Conservation Biology, Volume 27, Issue 1, Published on- 30 October 2012

Cumming, G. S.; Barnes, G.; Perz, S.; Schmink, M.; Sieving, K. E.; Southworth, J.; Binford, M.; Holt, R. D.; Stickler, C. and Holt, T. Van. (2005) *An Exploratory Framework for the Empirical Measurement of Resilience,* Ecosystems (2005) 8: 975–987 DOI: 10.1007/s10021-005-0129-z, Published online 28 November 2005.

Di Falco, Salvatore and Chavas, Jean-Paul (2008) *Rainfall Shocks, Resilience, and the Effects of Crop Biodiversity on Agroecosystem Productivity,* Land Economics N February 2008 N 84 (1): 83–96 ISSN 0023-7639; E-ISSN 1543-8325, Published on February 2008.

Doak, Daniel F.; Estes, James A.; Halpern, Benjamin S.; Jacob, Ute; Lindberg, David R.; Lovvorn, James; Monson, Daniel H.; Tinker, M. Timothy; Williams, Terrie M.; Wootton, J. Timothy; Carroll, Ian; Emmerson, Mark; Micheli, Fiorenza and Novak, Mark (2008) *Understanding and Predicting Ecological Dynamics: Are Major Surprises Inevitable,* Ecology, 89(4), 2008, pp. 952–961, 2008 by the Ecological Society of America, Published on- 2008

Dumanski, J., R. Peiretti, J. Benetis, D. McGarry, and C. Pieri (2006) *The paradigm of conservation tillage,* Proc. World Assoc. Soil and Water Conserv., P1: 58-64, Published on- 31 August,2006

Dumas, Sarah E.; Lungu, Luke; Mulambya, Nathan; Daka, Whiteson; McDonald, Erin; Steubing, Emily; Lewis, Tamika; Backel, Katherine; Jange, Jarra; Lucio-Martinez, Benjamin; Lewis, Dale; Travis, Alexander J. (2016) *Sustainable smallholder poultry interventions to promote food security and social, agricultural, and ecological resilience in the Luangwa Valley, Zambia,* Food Sec. (2016) 8:507–520 @ Springer Science+Business Media Dordrecht and International Society for Plant Pathology 2016, Published online: 1 June 2016

Du, Ling; McCarty, Gregory W.; Li, Xia, Rabenhorst, Martin C.; Wang, Qianfeng; Lee, Sangchul; Hinson, Audra L. and Zou, Zhenhua (2021) *Spatial extrapolation of topographic models for mapping soil organic carbon using local samples,* Geoderma 404 (2021) 115290, Elsevier, ScienceDirect, Published online 19 June 2021

Fabietti, Gabriele; Biasioli, Mattia; Barberis, Renzo and Ajmone-Marsan, Franco (2009) *Soil contamination by organic and inorganic pollutants at the regional scale: the case of Piedmont, Italy,* J Soils Sediments (2010) 10:290–300 DOI 10.1007/s11368-009-0114-9, Published online - 11 July 2009

Farooq, Muhammad and Siddique, Kadambot H. M. (2014) *Conservation Agriculture: Concepts, Brief History, and Impacts on Agricultural Systems,*Conservation Agriculture, Springer Champ, Published online: 04 December 2014

Fiksel, Joseph (2006) *Sustainability and resilience: toward a systems approach,* Sustainability and resilience: toward a systems approach, Sustainability: Science, Practice and Policy, 2:2, 14-21, DOI: 10.1080/15487733.2006.11907980, Published online: 05 Oct 2017

Fischer, Joern; Lindenmayer, David B. and Manning,Adrian D. (2006) *Biodiversity, ecosystem function, and resilience: ten guiding principles for commodity production landscapes,* Published on 2006

Flores, Bernardo M; Staal, Arie; C, Catarina; Jakovac; Hirota, Marina; Holmgren, Milena & Oliveira, Rafael S. (2019) *Soil erosion as a resilience drain in disturbed tropical forests*, Springer, Published online: 08 May 2019

Folke, Carl; Carpenter, Steve; Elmqvist, Thomas; Gunderson, Lance; Holling, C. S. and Walker, Brian (2002) *Resilience and Sustainable Development: Building Adaptive Capacity in a World of Transformations,* AMBIO: A Journal of the

Human Environment, 31(5):437-440. Published By: Royal Swedish Academy of Sciences https://doi.org/10.1579/0044-7447-31.5.437 URL: http://www.bioone.org/doi/full/10.1579/0044-7447-31.5.437, Published on 5 August 2002

Foster, David; Swanson, Frederick; Aber, John; Burke, Ingrid; Brokaw, Nicholas; Tilman, David and Knapp, Alan (2003) *The Importance of Land-Use Legacies to Ecology and Conservation*, BioScience, 53(1):77-88. 2003. Published by: American Institute of Biological Sciences, Published on- 1 January 2003

Franklin, Jerry F. (1989) *Importance and Justification of Long-Term Studies in Ecology*, Long-Term Studies in Ecology, Springer-Verlag New York Inc. 1989, Published on- 1989

Froyd, C.A. and Willis, K.J. (2008) *Emerging issues in biodiversity & conservation management: The need for a palaeoecological perspective*, Quaternary Science Reviews 27 (2008) 1723–1732, Published on- 2008

Garbach, Kelly; Milder, Jeffrey C.; DeClerck, Fabrice A.J.; Montenegro de Wit, Maywa; Driscoll, Laura and Gemmill-Herren, Barbara (2016) *Examining multi-functionality for crop yield and ecosystem services in five systems of agroecological intensification*, International Journal of Agricultural Sustainability ISSN: 1473-5903 (Print) 1747-762X, Published on- 28 April 2016

Ghosh, Sonaka; Das, T K; Sharma, D K and Gupta, Kamlika (2019) *Potential of conservation agriculture for ecosystem services: A review*, Indian Journal of Agricultural Sciences 89 (10): 1572–9, October 2019/Review Article

Giangrande, Adriana (2003) *Biodiversity, conservation, and the 'Taxonomic impediment'*, Aquatic Conservation: Marine and Freshwater Ecosystems. 13: 451–459 (2003), Published on- 2003

Giller, Ken E.; Andersson, Jens A; Corbeels, Marc; Kirkegaard, John; Mortensen, David; Erenstein, Olaf and Vanlauwe, Bernard (2015) *Beyond Conservation Agriculture*, frontiers in Plant Science, Agroecology, Published on- 28 October 2015

Greiner, Romy and Gregg, Daniel (2011) *Farmers' intrinsic motivations, barriers to the adoption of conservation practices and effectiveness of policy instruments: Empirical evidence from northern Australia*, Land Use Policy 28 (2011) 257–265, Elsevier Science, Published on- 2011

Grönwall, Jenny& OduroKwarteng, Sampson(2017) *Groundwater asastrategic resource forimproved resilience: acase study fromperiurban Accra*, Springer, Published online: 22 December 2017

Gunderson, Lance H. (2000) *Ecological Resilience- In Theory and Application*, Annual Review of Ecology and Systematics, Vol. 31:425–439, Publishing Year- November 2000

Hails, Rosemary S. and Ormerod, Steve J. (2013) *Ecological science for ecosystem services and the stewardship of Natural Capital,*Journal of Applied Ecology 2013, 50, 807–811, @British Ecological Society, Published on- 15 July 2013

Hughes, Terence P.; Bellwood, David R.; Folke, Carl; Steneck, Robert S. and Wilson,James (2005) *New Paradigms for Supporting the Resilience of Marine Ecosystems,* TRENDS in Ecology and Evolution Vol.20 No.7 July 2005, Elsevier Science, Published online 9 April 2005.

Hurni, Hans; Herweg, Karl; Portner, Brigitte; and Liniger, Hanspeter (2008)*Soil Erosion and Conservation in Global Agriculture,* Land Use and Soil Resources. 41 © Springer Science + Business Media B.V. 2008, Published on- 2008

Jackson, Louise; Noordwijk, Meine van; Bengtsson, Janne; Foster, William; Lipper, Leslie; Pulleman, Mirjam; Said, Mohammed; Snaddon, Jake and Vodouhe, Raymond (2010) *Biodiversity and agricultural sustainagility: from assessment to adaptive management,* Current Opinion in Environmental Sustainability 2010, 2:80–87, Elsevier Science, Published on 2010.

Kassam, A.; Derpsch, R. and Friedrich, T. (2014) *Global achievements in soil and water conservation: The case of Conservation Agriculture,* International Soil and Water Conservation Research, Vol. 2, No. 1, 2014, pp. 5-13, Published on- 2014

Kassam, A.; Friedrich, T. and Derpsch, R. (2018) *Global spread of Conservation Agriculture,* International Journal of Environmental Studies, https://doi.org/10.1080/00207233.2018.1494927, Published online- 06 Aug 2018

Katic, Pamela & Grafton, R. Quentin (2011) *Optimal groundwater extraction under uncertainty: Resilience versus economic payoffs,* Elsevier Science, Published online 7 July 2011

Khan, Zeyaur; Midega, Charles; Pittchar, Jimmy; Pickett, John and Bruce, Toby (2011) *Push- pull technology: a conservation agriculture approach for integrated management of insect pests, weeds and soil health in Africa,* International Journal of Agricultural Sustainability, 9:1, 162-170, Published online- 08 June 2011.

Kimmel, Courtney E. and Hull, R. Bruce (2012) *Ecological Entrepreneurship Support Networks: Roles and functions for conservation organizations,* Geoforum 43 (2012) 58–67, Elsevier Science, Published online- 30 August 2011

King, Christine A. (2008) *Community Resilience and Contemporary Agri-Ecological Systems: Reconnecting People and Food, and People with People,* Systems Research and Behavioral Science Syst. Res. 25, 111-124 (2008) Published online in Wiley InterScience

Knickel, Karlheinz (1990)*Agricultural Structural Change: Impact on the Rural Environment,* Journal of Rural Studies, Vol. 6, No. 4, pp. 383-393, 1990, @ researchgate

Knowler, Duncan &Bradshaw,Ben (2007) *Farmers' adoption of conservation agriculture: A review and synthesis of recent research,* Elsevier Science, Published on- 2007

Kosmas, Costas; Karamesouti, Mina; Kounalaki, Kate; Detsis, Vassilis; Vassiliou, Penny; Salvati, Luca (2016*) Land degradation and long-term changes in agro-pastoral systems: An empirical analysis of ecological resilience in Asteroussia - Crete (Greece),* Elsevier Science, Published online 15 July 2016

Kumaraswamy, S. and Kunte, K. (2013) Integrating biodiversity and conservation with modern agricultural landscapes, Biodivers Conserv (2013) 22:2735–2750, Published online- 19 September 2013

Lal, Rattan (2015) *Sequestering carbon and increasing productivity by conservation agriculture,* Journal of Soil and Water Conservation May/June 2015—VOL. 70, NO. 3, Published on- 2015

Lal,R. (1997) *Degradation and resilience of soils,* The Royal Society, Published on 1997.

Lambooy, Tineke and Levashova, Yulia (2011) *Opportunities and challenges for private sector entrepreneurship and investment in biodiversity, ecosystem services and nature conservation,* International Journal of Biodiversity Science, Ecosystem Services & Management Vol. 7, No. 4, December 2011, 301–318, Published online- 13 Jan 2012.

Lindenmayer, David B.; Likens, Gene E.; Andersen, Alan; Bowman, David; Bull, C. Michael; Burns, Emma; Dickman, Chris R.; Hoffmann, Ary A.; Keith, David A.; Liddell, Michael J.; Lowe, Andrew J.; Metcalfe, Daniel J.; Phinn, Stuart R.; Russell-Smith, Jeremy; Thurgate, Nikki and Wardle, Glenda M. (2012) *Value of long-term ecological studies,* Published on- Ecological Society of Australia, 2012

Lindenmayer, D.B. (1999) *Future directions for biodiversity conservation in managed forests: indicator species, impact studies and monitoring programs,* Forest Ecology and Management 115 (1999) 277-287, Elsevier Science, Published on-1999

Maggioni, Elena (2015) *Water demand management in times of drought: What matters for water conservation,* Water Resources Research 10.1002/2014WR016301, Published online- 8 January 2015

Mathevet,Raphaël; Bousquet, François; Larrère, Catherineand Larrère, Raphaël(2018) *Environmental Stewardship andEcological Solidarity: Rethinking SocialEcological Interdependency andResponsibility,*Journal of Agricultural and Environmental Ethicsvolume**31**,pages605–623(2018), @Springer, Published on- 28 October 2018

Mcneely, Jeffrey A. and Schroth, Go¨Tz (2006)*Agroforestry and biodiversity conservation – traditional practices, present dynamics, and lessons for the future,* Biodiversity and Conservation (2006) 15:549–554, Springer 2006

Moller, Henrik; J. MacLeod, Catriona; Haggerty, Julia; Rosin, Chris; Blackwell, Grant; Perley, Chris; Meadows, Sarah; Weller, Florian & Gradwohl, Markus (2008) *Intensification of New Zealand agriculture: Implications for biodiversity*, New Zealand Journal of Agricultural Research, 51:3, 253-263, DOI: 10.1080/00288230809510453, Published online: 15 July 2008

Moss, Brian (2007)*Water pollution by agriculture*, Philosophical Transactions of the Royal Society B 363:659-666, Published on- 30 July 2007

Mwase, Weston; Sefasi, Abel; Njoloma, Joyce; Nyoka, Betserai I.; Manduwa, Daniel and Nyaika, Jacinta (2015) *Factors Affecting Adoption of Agroforestry and Evergreen Agriculture in Southern Africa*, Environment and Natural Resources Research; Vol. 5, No. 2; 2015 ISSN 1927-0488 E-ISSN 1927-0496, Published online- April 25, 2015.

Nearing, Mark A.; Xie, Yun; Liu, Baoyuan and Ye, Yu (2017) *Natural and anthropogenic rates of soil erosion*, International Soil and Water Conservation Research ,Volume 5, Issue 2,June 2017, Pages 77-84, @ScienceDirect Elsevier, Published on- 18 April 2017

Nelson, Erik; Mendoza, Guillermo; Regetz, James; Polasky, Stephen; Tallis, Heather; Cameron, D Richard; Chan, Kai MA; Daily, Gretchen C; Goldstein, Joshua; Kareiva, Peter M; Lonsdorf, Eric; Naidoo, Robin; Ricketts, Taylor H and Shaw, M Rebecca (2009) *Modeling multiple ecosystem services, biodiversity conservation, commodity production, and tradeoffs at landscape scales*, The Ecological Society of America, Frontiers in Ecology and the Environment, Volume18, Issue 10, Online ISSN:1540-9309, Published on-01 February 2009

Nichols, James D.; Koneff, Mark D.; Heglund, Patricia J.; Knutson, Melinda G.; Seamans, Mark E.; Lyons James E.; Morton, John M.; Jones, Malcom T.; Boomer, G. Scott and Williams, Byron K. (2011) *Climate change, uncertainty, and natural resource management*, The Journal of Wildlife Management, Volume 75, Issue 1, Pages- 6-18, Published on- 31 January 2011

Oliver, Tom H.; Heard, Matthew S.; Isaac, Nick J.B.; Roy, David B.; Procter, Deborah; Eigenbrod, Felix; Freckleton, Rob; Hector, Andy; Orme, C. David L.; Petchey, Owen L.; Proença, Vânia; Raffaelli, David; Suttle, K. Blake; Mace, Georgina M.; Martín-López, Berta; Woodcock, Ben A. and Bullock,James M. (2015), *Biodiversity and Resilience of Ecosystem Functions*, Trends in Ecology & Evolution, November 2015, Vol. 30, No. 11, Published on November 2015.

Olsson, Per and Folke, Carl (2001) *Local Ecological Knowledge and Institutional Dynamics for Ecosystem Management: A Study of Lake Racken Watershed, Sweden*, Ecosystems · March 2001 DOI: 10.1007/s100210000061, Published on March 2001

Palma, Cheryl; Blanco-Canqui, Humberto; DeClerck, Fabrice; Gaterea, Lydiah and Grace, Peter (2014) *Conservation agriculture and ecosystem services: An overview*,Agriculture, Ecosystems & Environment, Volume 187,1 April 2014, Pages 87-105

Peterson, Garry; Allen, Craig R. and Holling, C. S. (1998) *Ecological Resilience, Biodiversity, and Scale*, Ecosystemsvolume**1**,pages 6–18(1998)

Perrings, Charles and Walker,Brian (1997) *Analysis Biodiversity, resilience and the control of ecological-economic systems: the case of fire-driven rangelands*, Elsevier Science B.V., Published on 1997

Perry-Hill, R. and Prokopy, L.S. (2014) *Comparing different types of rural landowners: Implications for conservation practice adoption*, Journal of Soil and Water Conservation: May/June 2014—Vol. 69, No. 3, Published on- May/June 2014.

Pittelkow, Cameron M.; Liang, Xinqiang; Linquist, Bruce A.; Groenigen, Kees Jan van; Lee, Juhwan; Lundy, Mark E.; Gestel, Natasja van; Six, Johan; Venterea, Rodney T. and Kessel, Chris van (2014) *Productivity limits and potentials of the principles of conservation agriculture*, NATURE | Vol 000 | 00 Month 2014

Power, Alison G. (2010) *Ecosystem services and agriculture: tradeoffs and synergies*,Philosophical Transactions of the Royal Society B Biological Sciences, Volume365 Issue 1554, PubMed: 20713396, Published by: Royal Society, Print ISSN:0962-8436, Online ISSN:1471-2970, Published on- 16 August 2010

Pradhan, Aliza; Chan, Catherine; Roul, Pravat Kumar; Halbrendt, Jacqueline; Sipes, Brent (2017) *Potential of conservation agriculture (CA) for climate change adaptation and food security under rainfed uplands of India: A transdisciplinary approach*, Agricultural Systems (2017), http://dx.doi.org/10.1016/j.agsy.2017.01.002, Published on-2017

Prager, K.; Schuler, J.; Helming, K.; Zander, P.; Ratinger, T. and Hagedorn, K. (2011) *Soil Degradation, Farming Practices, Institutions and Policy Responses: An Analytical Framework*, Land Degradation & Development, 22: 32–46 (2011), Published on-2011

Ray, G.L.; Mondal, Sagar (2011) *Research Methods in Social Sciences and Extension Education*, Kalyani Publishers, Published on-2011

Rescia, Alejandro J.; Willaarts, Bárbara A.; Schmitz, María F.; Aguilera,Pedro A. (2010) *Changes in land uses and management in two Nature Reserves in Spain: Evaluating the social-ecological resilience of cultural landscapes*, Landscape and Urban Planning, Elsevier Science, Published online 21 August 2010.

Roesch-McNally, Gabrielle E.; Basche, Andrea D.; Arbuckle, J.G.; Tyndall, John C.; Miguez, Fernando E.; Bowman, Troy and Clay, Rebecca (2017) *The trouble with cover crops: Farmers' experiences with overcoming barriers to adoption*, Renewable Agriculture and Food Systems: doi:10.1017/S1742170517000096, Published on- 2017.

Sanderson, Matt A.; Archer, David; Hendrickson, John; Kronberg, Scott; Liebig, Mark; Nichols, Kris; Schmer, Marty; Tanaka, Don and Aguilar, Jonathan (2013) *Diversification and ecosystem services for conservation agriculture: Outcomes from pastures and integrated crop–livestock systems,* Renewable Agriculture and Food Systems, Volume 28, Issue 2, June 2013, pp. 129 – 144, Published on- 11 March 2013

Sarangi, A.; Madramootoo, C. A. and Cox, C. (2004) *A Decision Support System for Soil and Water Conservation Measures on Agricultural Watersheds,* Land Degradation and Development, 15: 49–63 (2004), Published on- 2004

Saysel, Ali Kerem; Barlas, Yaman and Yenigun, Orhan (2002) *Environmental sustainability in an agricultural development project: a system dynamics approach,* Journal of Environmental Management, Volume 64, Issue 3, Pages 247-260, @ ScienceDirect Elsevier, Published on- March 2002

Seidl, Irmi; Schelske, Oliver and Joshi, Jasmin (2003) *Entrepreneurship in biodiversity conservation and regional development,* Entrepreneurship & Regional Development, 15, October–DECEMBER (2003), 333–350, Published online- 2003.

Shava, Soul; Krasny, Marianne E.; Tidball, Keith G. and Zazu, Cryton (2010) *Agricultural knowledge in urban and resettled communities: applications to social–ecological resilience and environmental education,* Environmental Education Research Vol. 16, Nos. 5–6, October–December 2010, 575–589

Singh, SP; Naresh, RK; Panwar, AS; Vivek; Mahajan, NC; Kumar, Sudhir; Tomar, SK; Mrunalini, Kancheti and Prasad, KS Krishna (2019) *Conservation agriculture: Building entrepreneurship and resilient farming systems: A review,* International Journal of Chemical Studies 2019; 7(2): 990-997, Published on- April 2019

Su, Yang; Gabrielle, Benoit and Makowski, David (2021) *A global dataset for crop production under conventional tillage and no tillage systems, Sci Data***8**,33 (2021). https://doi.org/10.1038/s41597-021-00817-x, Published on 28 January 2021

Teague, W.R.; Apfelbaum, S.; Lal, R.; Kreuter, U.P.; Rowntree, J.; Davies, C.A.; Conser, R.; Rasmussen, M.; Hatfield, J.; Wang, T.; Wang, F.; and Byck, P. (2016) *The role of ruminants in reducing agriculture's carbon footprint in North America,* Journal of Soil and Water Conservation, Published on- April 2016

Thierfelder, C. and Wall, P. C. (2010) *Rotation in Conservation Agriculture Systems of Zambia: Effects on Soil Quality and Water Relations,* Expl Agric. (2010), volume 46 (3), pp. 309–325, doi:10.1017/S001447971000030X, Published online- 9 June 2010

Vatn, Arild; Bakken, Lars; Bleken, Marina A.; Baadshaug, Ole Hans; Fykse, Haldor; Haugen, Lars E.; Lundekvam, Helge; Morken, John; Romstad, Eirik; Rørstad, Per Kristian; Skjelva˚g, Arne O.; Sogn, Trine (2006) *A methodology for integrated economic and environmental analysis of pollution from agriculture,* Agricultural Systems, Volume 88, Issues 2–3, Pages 270-293, @ScienceDirect Elsevier, Published on- June 2006

Walker, Brian; Holling, C.S; Carpenter, Stephen R. and Kinzig, Ann (2004) *Resilience, Adaptability and Transformability in Social–ecological Systems,* Ecology and Society, Dec 2004, Vol. 9, No. 2 (Dec 2004), Published by: Resilience Alliance Inc.

Whitlow, Richard (1988) *Soil erosion and conservation policy in Zimbabwe: Past, present and future,* Land Use Policy, Volume 5, Issue 4, Pages 419-433, @ScienceDirect, Elsevier, Published on- October 1988

Willis, Katherine J; Araújo, Miguel B; Bennett, Keith D; Figueroa-Rangel, Blanca; Froyd, Cynthia A and Myers, Norman (2007) *How can a knowledge of the past help to conserve the future? Biodiversity conservation and the relevance of long-term ecological studies,* Phil. Trans. R. Soc. B 2007 362, doi: 10.1098/rstb.2006.1977, Published on- 28 February 2007

Willis, K. J. and Birks, H. J. B. (2006) *What Is Natural? The Need for a Long-Term Perspective in Biodiversity Conservation,* Science Vol 314 24 November 2006, Published on- 24 November 2006

Wilkinson, Cathy (2011) *Social-ecological resilience: Insights and issues for planning theory,* SAGE journalsISSN: 1473-0952 Online ISSN: 1741-3052, Published on- November 6, 2011

Winn, Monika I. and Pogutz,Stefano (2013) *Business, Ecosystems, and Biodiversity: New Horizons for Management Research,* Published on February 10, 2015.

Worrell, Richard and Appleby, Michael C. (2000) *Stewardship of Natural Resources: Definition, Ethical and Practical Aspects,* Journal of Agricultural and Environmental Ethics 12, pages 263-277(2000),@Springer, Published on- January 2000

Worstell, Jim (2020) *Ecological resilience of food systems in response to the COVID-19 crisis,* Journal of Agriculture, Food Systems, and Community Development, 9(3), 23–30, Published online April 19, 202

Part- II
Entrepreneurial Growth and Communication Behavior in Conservation Agriculture: The Role and Response of Farmer

Synopsis

A good quality land yields good results to everyone, confers good health on the entire family, and causes growth of money, cattle and grain. As a source of livelihood, agriculture remains the largest sector of Indian Economy. While its output share fell from 28.3% in 1993-94 to 14.4% in 2011- 12 and employment share declined from 64.8% to 48.9% over the same period. Therefore, almost half of the workforce in India still remains dependent on agriculture. Given the low share of this workforce in the GDP, on average, it earns much lower income poorer than its counterpart in industry and services. Agriculture faces many challenges, making it more and more difficult to achieve its primary objective - feeding the world – each year. By 2050 the world's population will reach 9.8 billion, 34 percent higher than today. Annual cereal production will need to rise to about 3 billion tonnes from 2.1 billion today and annual meat production will need to rise by over 200 million tonnes to reach 470 million tonnes. Agriculture produces an average of 23.7 million tonnes of food every day. To provide for a population of 9.8 billion in 2050, and 11.2 billion in 2100, food production will need to increase from the current 8.4 billion tonnes to almost 13.5 billion tonnes a year. Fifty percent of the additional food required to meet demand in 2050 will need to come from land already under cultivation. Seventy-five percent of the genetic diversity of crops has already been lost. By 2050, two out of three people on the planet will live in urban areas; a large portion of future urbanization will be caused by rural-urban migration (Buhaug and Urdal, 2013). CA addresses a wide range of agricultural production challenges that include declining soil fertility, increasing production costs; climate induced erratic rainfall patterns and increased demand for food production against severely reduced production capacities of agricultural lands. Solutions are workable options that can be tailored to raising system productivity or diversity, efficiency, resilience, value and profitability of farming, including the enabling mechanisms needed within diverse local contexts. Advances towards building entrepreneurship and resilient farming systems are the most effective and durable strategies, where all stakeholders work together to bring their ideas and support to developing and implementing site-specific solutions that allow for iterative, continuous improvement of the world's food systems and their key components. The paper offers ideas on how these problems can be addressed so as to accelerate agricultural growth in a sustainable manner.

Key words: *Conservation, Entrepreneurial behaviour, Ecological services, Enterprise perception.*

7

Introduction

Smallholder agriculture is the mainstay of food production in the world's developing countries and is the key to ensuring long-term global food security (FAO, 2014). With the global population on course to exceed 9 billion by 2050, the need to meet the growing demand for more food is immediate and pressing. The world's 500 million smallholder farms currently produce around 80 percent of our food and it is they who will have to bear the brunt of the need to increase food production by over 60 percent compared to 2007 levels. Currently many of these smallholder farms have limited access to production inputs and so achieve low levels of productivity. One key production input i.e. mechanization, is frequently neglected in farm productivity improvement efforts; in fact it has been described as the neglected waif of agricultural and rural development. Increasing food production whilst conserving the planet's natural resource base will not be a simple task. In second Green Revolution, which aimed at doubling of global food production in the second half of last century, became implausible. Rates of growth in the yields of the world's major food cereals (wheat, rice and maize) are now falling and this is due in no small part to the degradation of agricultural land.

Increase in food production via a process of sustainable intensification will necessarily, require the implementation of more natural resource-friendly production methods, for example reduced- and no-till farming, and this will require a major diffusion of novel mechanization technology. Cropping intensity can be raised in a variety of ways as a result of adopting conservation agriculture (CA) system. Plough-based tillage is intensely time and energy-consuming. Primary tillage can also only be easily accomplished when soil conditions are suitable-not too dry and not too wet. Waiting until the first rains before initiating soil preparation can waste valuable time, delay planting until late in the sowing window, and results in reduced yields. Although highly variable, yield losses of up to one percent per day of delay after the optimum planting date can be experienced. Yield increases can also be expected after switching to CA (Ngwira et al., 2013) but improvements cannot always be expected to be immediate. Degraded soils will need time to recover their health, structure and fertility under a no-till regime so, some yield decline may be experienced before yields recover and exceed pre-switchover levels. Integrating livestock production with CA cropping is another possibility. Using leguminous cover crops to feed a livestock enterprise such as dairy, goats could be a possibility if cultural preferences and market conditions are appropriate. A current phenomenon affecting many rural economies in the developing world is the drift of healthy young males to urban centers in search of more rewarding payment for their efforts. Fifty percent of the world's global population is urban today and this is projected to rise to 70 percent by 2050 (FAO, 2011). This means that those being left behind to work on the farms are women, the elderly and children and the consequence is that farm power becomes an increasingly severe constraint. If the supply of human labor emanates principally from women, the elderly and children, it is clear that supply constraints will have a negative impact on farm productivity. Improving smallholders' access to crucial mechanization inputs is frequently fraught with difficulties as the adoption of any innovation must be seen as a useful and profitable investment from the perspective of the farmer. This prerequisite would then trigger the necessary demand for these innovations which should lead to an increased or stabilized supply. However, very often this demand is nonexistent due to the low income levels of small farmers which in turn also lead to only very rudimentarily developed mechanization input supply chains in rural areas. This situation is referred to as 'the vicious cycle of mechanization development'. It requires broad action among all actors involved in rural development and sustainable intensification programs from the farmers' level up to and including policies to disintegrate the cycle (Kienzle and Sims, 2014). Mechanization inputs are usually lumpy so that a smallholder with, typically, under two hectares of land, will be reluctant to be the sole investor in a machine which has the potential to operate over a much larger area. As always, there will be many demands on a farm families financial assets and the opportunity cost of capital may militate against adoption of machinery. The affordability of mechanization inputs in the developing country smallholder farming sector is closely correlated with their profitability. Almost by definition poor smallholder families struggling to emerge from subsistence farming will usually have difficulty in amassing the resources necessary for machinery purchase. An alternative is to borrow the money required but this can be expensive and formal sources of credit, such as banks, are notoriously reluctant to extend credit to the sector, citing the high level of risk involved. Where specialist equipment is required for the adoption of sustainable crop intensification by means of conservation agriculture, then a further obstacle to adoption arises in the form of low availability. Supply is frequently initiated by externally funded aid programs and projects and local supply chains are generally not adequately developed. Machinery is often donated or sold at cost under attractive loan arrangements which are typified by low interest rates, long payback periods and a benign attitude towards defaulters. This can lead to the situation where farmers outside the programmed target group are unable to access the machinery, even if they are convinced that it would be beneficial, as there is no local supply network. This situation is slowly improving as donor organizations realize the value of involving local dealers in the supply of machinery. Purchasing power of small and marginal farmers can be improved through farmer groups and Farmer Producer Organizations (FPOs). This aggregation approach also helps small and marginal farmers in accessing various benefits of government schemes for rural development. CA, in conjunction with good crop, nutrient, weed and water management, is at the heart of FAO's new sustainable agricultural intensification strategy (FAO, 2011) which takes an ecosystems approach to enhance productivity and resilience as well as the flow of ecosystem services while reducing emissions that come from the agriculture sector (Kassam et al., 2011a). These characteristics are also an integral part of climate-smart agriculture that seeks to increase productivity in an environmentally and socially sustainable way, strengthen farmers' resilience to climate change, and reduce GHG emissions and sequester carbon (World Bank, 2012). At the heart of sustainable agricultural intensification, or sustainable land management, is the integration of soil and water conservation practices in agricultural production, with concurrent objectives of enhanced economic returns and environmental management. Under no till, a much richer and more favourable soil biological environment is created in the soil, promoting larger amounts and diversity of microorganisms and soil meso and macro-fauna. These generate and control some of the ecosystem functions critical for good soil health, including soil carbon storage and nutrient cycling. They are also important in promoting larger and more stable soil aggregates (Wright, 1998), as well as networks of soil "bio-pores", thereby promoting improved water infiltration and soil water storage. The extensive cover of the soil surface by organic residues and stubble greatly reduces the amount and severity of water run-off and soil erosion. In turn, this reduces surface water pollution from the sediments and solutes that are regularly carried with the eroded soil. The surface litter associated with no tillage, as well as the increased soil organic matter, greatly enhance the capacity of the soil to capture and store rain and irrigation water. At the same time, it significantly reduces surface water runoff, soil erosion, and evaporation. These changes ensure at least some level of drought proofing during dry periods, and they help to reduce yield variability among seasons, thereby facilitating better farm planning. Under irrigated conditions, they significantly reduce the amount of water needed to bring a given crop to maturity. New technologies will make it possible for conservation agriculture to building entrepreneurship and resilient farming systems to become the new global standard, not the exception; the main factors resisting change are political will, lack of policy coherence at many levels, financing, governance and human behaviour. We propose evidence-based indicators that could be applied to track progress towards meeting the entrepreneurship and resilient farming systems and their targets, at local, national, regional and global scales. Their effective use will require investing more in monitoring agriculture and food systems, taking advantage of rapid advances in digital information technologies. The transformation of agriculture will also require re-thinking of international and national structures. The global food system should morph into a true global partnership that widely shares information, experiences and new technology, following open access principles and practices that honor intellectual property but enable wide access and use. New models for implementation are needed that unlock the real potential of farmers, public and private sectors in solving complex problems. The private sector will be a key player in entrepreneurship and resilient farming systems and food systems. Good governance will be essential, including supporting farmer groups, managing risks, and deploying tools and accountability measures that foster greater private sector investment in agriculture, but also put clear constraints on unsustainable or inequitable exploitation of land, water, forests and fisheries.

Agriculture ecosystem services are usually called provisioning services, and have been regarded as the single most important role of farming (Scherr and McNeely, 2008). Farmers influence the capacity of the farming system to deliver ecosystem services. But farms not only sustain provisioning services but also provide supporting, regulating and cultural services. Supporting and regulating ecosystem services are mostly biologically regulated. Soil management is a key factor in many of the provisioning and supporting ecosystem services provided by farming systems. Maintaining a healthy soil is an important strategy for many supporting and regulating services: water infiltration and retention, resistance to erosion, nutrient cycling and retention, and disease suppressiveness (Powlson et al., 2011). Farmers actively manage soil health to secure future production, by enhancing regulating and supporting services of soils. Moreover, in the process of agricultural intensification, farmers have replaced much of the regulating and supporting ecosystem services that originally were provided for by soil biota, by human external inputs. Decomposing functions of soil organisms have been replaced by inorganic fertilizers, disease suppressive functions of soil organisms and natural biodiversity have been replaced by chemical crop protection agents and the function of 'ecosystem engineers' like earthworms in building soil structure has been diminished by tillage. (Giller et al., 1997). Changing water flow and decreased storage: Agriculture modifies the plant species composition and below-ground root structure, the production of litter, the extent and timing of plant cover, and the composition of the soil biotic community: factors that influence to a major extent are water infiltration and retention in soil. The intensity of agricultural production and management practices affect both the quantity and quality of water in agricultural landscapes (Power, 2010). Increasing erosion: Soil erosion is a natural geological phenomenon that has created the vast fertile soils of alluvial flood plains and loess plateaus around the world. However, the accelerated soil erosion, exacerbated by human perturbations, is a destructive process. It depletes soil fertility, degrades soil structure and reduces the rooting depth of plants (Lal, 2003). As a result, the diversity of plants, animals and microorganisms is diminished. Ultimately, the resilience of the entire ecosystem is threatened (Pimentel and Kounang, 1998). Land which is covered by plant biomass, living or dead, is protected and will experience reduced soil erosion. Vegetation is the main component of ecosystem biomass, followed by belowground microbial biomass. Maintenance of vegetation is one of the main principles to prevent soil erosion. In forested areas, a minimum of 60% forest cover of the landscape is necessary to prevent soil erosion and landslides. Another factor influencing the susceptibility to erosion is soil structure. Soils with medium to fine texture, low organic matter content, and weak structural development are most easily eroded. Principles for preventing soil erosion should enhance soil organic matter contents (Pimentel and Kounang, 1998). Changing climate: The global release of soil organic carbon from agricultural activities has been estimated at 800 Tg C yr1 (T = tera = 1012) Soil biological degradation, by decrease of soil organic carbon, is an important factor leading to C emission from soil to atmosphere (Lal, 1997). Increased CO2 represents the most important human enhancement to the greenhouse effect. Changes in land use and land cover, mainly driven by agriculture, may also influence climate through changes in evapo-transpiration. There is increasing concern about potential effects of land cover changes on agriculture, through adverse effects on the monsoons (Gordon et al., 2010).

With this background the present study has been designed and directed with the following rationale:

I. We need to have a paradigm shift from conservation to stewardship approaches so that the natural resource conserved can go in a resilient manner to support income, livelihood and social progress in interactive dispositions.

II. Our farmers are mostly input concerned and less energy conscious and the least concerned on stewardship aspect.

III. The soil, water and biodiversity are the prime movers to usher the transformation process of conservation approaches into stewardship functions and for which we need to generate baselines as well as preparedness of the subject farmers as to how and where they stand for stewardship functions.

IV. The stewardship functions for this empirical study can be estimated through different farm operation, management, indigenous knowledge, praxis, community innovation and social networking.

V. Whether the issues can be redressed in terms of non-cognitive stewardship with cognitive stewardship; functional stewardship with both non cognitive and cognitive, and ecological deliverables with stewardship functions and thereafter a disposition of dynamic interplay between ecology and sociology.

Hence the objectives are as follows:

I. To conduct a baseline survey for the research locale in order to identify appropriate respondents for the study.

II. To identify appropriate variables suitable for measuring the perception of the respondents towards conventional conservation practices in the study area.

III. To estimate the inter- and intra-level relationship among and between the set of variables to finally elucidate the perception of conservation practices.

IV. To assess the perception of the respondents on conventional conservation practices as well as possibility of stewardship.

V. To formulate extension intervention based on results.

8

Theory & Concepts

Conceptualizing the dynamics of farm stewardship based on conservation This chapter deals with the broadly described conceptual framework used in this study in order to provide a theoretical base for the empirical investigation and guidance for selection of relevant predictor variables.

8.1 Conceptual Framework

A conceptual framework is generally constructed to provide the researcher the elements of attainable and researchable problems and ideas supported by methodology, design, prospective analysis, and empirical conclusion. The topic of the present research being Paradigm Shift of Small and Marginal Farmers from Conventional Practices To Conservation practices to generate entrepreneurship: The Perception And Possibilities. It is expected that the terminologies and concepts used to frame up the topic need to undergo an analytical text and literature.

There has been a basic relation between conservation agriculture and Entrepreneurship. The former is revolving around maintenance to protect the count of resource and diversity, the latter keeps focusing on to be successful at market-oriented farming. The present conceptual framework has to accommodate and assimilate the idea of conserving nature with a sustainable growth in opportunities, employment and income generation in rural farm economy. And all these have to be configured within a conceptual framework the research is designing for.

Table 8.1: General principles related to different concepts of building entrepreneurship and resilient farming systems.

Building entrepreneurship and resilient farming systems concepts	Principles	References
Diversified Farming Systems	Farming practices and landscapes that intentionally include functional biodiversity at multiple spatial and/or temporal scales in order to maintain ecosystem services that provide critical inputs to agriculture, such as soil fertility, pest and disease control, water use efficiency, and pollination	Kremen et al., 2012

Eco-agriculture Landscapes	Mosaics of areas in natural/native habitat and areas under agricultural production. Agriculture, biodiversity and ecosystem services are seen as interdependent. Rural communities are critical stewards of biodiversity and ecosystem services.	Scherr and McNeely, 2008
Sustainable Intensification	Intensification of resources, using natural, social and human capital assets, combined with the use of best available technologies and inputs (best genotypes and best ecological management) that minimize harm to the environment. Resource-conserving technologies are Integrated Pest Management, Integrated Nutrient Management, Conservation Tillage, Agroforestry etc.	Pretty, 2008
Conservation Agriculture	Conservation Agriculture is defined as minimal soil disturbance (no-till, NT) and permanent soil cover (mulch) combined with rotations	Hobbs et al., 2008
Organic Agriculture	Organic agriculture is a production system that sustains the health of soils, ecosystems and people. It relies on ecological processes, biodiversity and cycles adapted to local conditions, rather than the use of inputs with adverse effects. Organic agriculture combines tradition, innovation and science to benefit the shared environment and promote fair relationships and a good quality of life for all involved.	IFOAM (www.ifoam.org)
Direct-seeding Mulch-based Cropping Systems	A form of Conservation Agriculture, Direct seeding mulch-based cropping systems are characterised by ideally no tillage, soil surface protection through mulch from crop residue, crop rotation, keeping the soil permanently covered	Affholder et al., 2010
Evergreen Agriculture	Evergreen agriculture is defined as the integration of particular tree species into annual food crop systems. The intercropped trees sustain a green cover on the land throughout the year to maintain vegetative soil cover, bolster nutrient supply through nitrogen fixation and nutrient cycling, generate greater quantities of organic matter in soil surface residues, improve soil structure and water infiltration, increase greater direct production of food, fodder, fuel, fiber and income from products produced by the intercropped trees, enhance carbon storage both aboveground and below-ground, and induce more effective conservation of above and below-ground biodiversity.	Garrity et al., 2010

8.2 Sustainability Concerns and Sustainable Entrepreneurship in Agriculture

The definition of sustainability is the general agreement to retain a balance between economic, environmental, and social factors on an equal and harmonious basis to meet current needs without affecting the needs of future generations. According to reports of the World Commission on Environment Development (WCED) (initially in 1934), in 1987, Schumpeter elaborated on the innovation process of creative destruction, whether it be climate change, biodiversity, or water shortages, in which entrepreneurs are constantly innovating based on environmental challenges. These degradations have indeed become a fuse for the entrepreneurial creativity of the future. Subsequently, in the 1970s, Harvard Business Review became a pioneer in the ecological movement, arguing that innovative businesses were the prospective solutions for the environment. Furthermore, in the 1990s, more research developed on eco-entrepreneurship, green entrepreneurship, and environmentally friendly business models. The entrepreneurial model upgraded to the double bottom line, namely eco-entrepreneurship. Further studies were done corresponding to this model, which indicated that maintaining future sustainability requires creativity, commitment to nature conservation, and a socio-economic culture or local awareness of today's environmental dynamics; in addition, entrepreneurship is improving and progressing gradually towards sustainable entrepreneurial development. Understanding sustainable entrepreneurship can be defined as individuals' recognition, development, and exploitation of opportunities to provide economic, social, and ecological benefits for future goods and services. Meanwhile, sustainable development can be defined as meeting the needs of the future which is supposed to be deprived of yielding the capability of future generation's needs, combining sustainable development plans with business models, and developing into synergizing continuous entrepreneurship. In 1987, the Brundtland Commission developed the concept of maintaining the progress of economic development without reducing natural resources or damaging the environment. In the years since, numerous sustainability statements were implemented in the field of entrepreneurial development. The simple idea of the model is to consider the triple bottom line approach. With the existence of sustainable entrepreneurship controls, the triple bottom line role of entrepreneurs is to manage the balance between the planet, humans, and profits. As sustainable entrepreneurship studies have progressed and as research continues to deepen, sustainable entrepreneurship has become a synergistic combination of entrepreneurial programs and sustainable development. Consideration of the principles of sustainable development leads to the use of natural resources that effects the balance between profit, social goals, and ecological goals. The scarcity of natural resources can have been endorsed in sustainable development. Singh created a mindset that has changed from a traditional model to a sustainable tomorrow's market, similar to sustainable entrepreneurship that claims to combine economic, social, and environmental value creation of sustainable agribusiness. While it was previously thought that entrepreneurs focus primarily on economic value creation, in these new entrepreneurship forms, economic value creation is seen as a means to an end or the blending of diferent values; implementation of environmental sustainability ideas has led to the sustainable development of enterprises and competitive advantage of each enterprise. In the view of Mc Ewe and Tilley and Young, sustainable entrepreneurship is a business model that integrates three domains of sustainable development. In short, the definition of sustainable entrepreneurship is not only limited to the creation of products and services, but also includes commitment and environmental equality. According to the convergence process model developed by Belz and Binder, business positioning is first taken from double bottom line stages and developed to triple bottom lines. Based on a report from Choi and Gray, it is pertinent that sustainable entrepreneurship creates profitable businesses and achieves specific environmental and social goals. Consequently, the incorporation of technical learning areas creates sustainability value for entrepreneur. In that spirit, sustainable entrepreneurs serve as economic agents to initiate, adapt, and integrate those processes and activities which lead to the development of sustainability-driven and profitable economic opportunities. Therefore, business innovation must ensure the sustainability of benefits without neglecting the needs of future generations

Expected outcomes of the study:

- The present status of conservation practices followed and perceived entrepreneurial possibilities among the farmer.
- The indicator to measure the paradigm shift from exploitative agriculture to conservational agriculture in terms of selected variables.
- The clandestine shift of practicing conservation to possible stewardship in terms of technology, resources and ecology.
- An extension Policy generation, as applicable for the research locale, as well as similar ecological set up elsewhere for integrating conservation with entrepreneurship through an approach of ecological resilience and economic approaches.

Research Question:

- Are the farmers really concerned about conservation, and if so what are the cognitive, perceptual and operational levels?
- What are the set of factors influencing and characterizing the nature of conventional conservational practices followed and serendipitous possibilities of entrepreneurship?
- Is there any paradigm shift from exploitative agriculture to conservative agriculture leaving a space for entrepreneurial possibilities to dovetail

productive farming with ecological resilience and employment generation?

- ❑ What are the extension and policy implication of the study?

Specific objectives:

This may be treated as the operational definition for farm conservation; else it's a conceptual and empirical improvisation over classical conventional strategy. The basic component of conservation for Indian farms can be derived in the following ways:

I. Base line information on input application, management practices and production status for performing farm and their interrelationship with ambience institutions.

II. Optimization of cost intensive chemical inputs in favor of organic input driven production process.

III. To estimate the sub ecological interactions and relations of the farms so that integrated and resilient farm production process can be ushered.

IV. The social compatibility, economic profitability and ecological suitability need to converge on a common agendum of deliverables with design and impact.

The application of the concept in the present empirical research has been proposed to follow both the inductive and deductive approaches. The approaches are as follows:

I. The study on connectivity between conservation of soil, water, biodiversity in terms of interdependent and interaction. these will follow application of both structure schedule and participatory approach

II. The study on connectivity between conservation of soil, water, biodiversity in terms of interdependent and interaction. These will follow application of both structure schedule and participatory approach.

III. To derive the possibility of applying various approaches through necessary transformation, up gradation and acculturation of conservation strategies which are being followed mostly by the practicing farmers.

IV. Identification of conceptual motivational and institutional constraints acting as a barrier to this desired transformation.

V. Through participatory designing and analysis, we can identify the people – ready status for ushering farm stewardship in terms of economic viability, social compatibility and ecological resilience.

9

Review and Citations

This chapter reviews the research work done in the fields related to the objectives of the study. In any scientific investigation or research a comprehensive review of relevant literature is very essential. Its main function apart from determining the work done before concerning the problem area i.e. area of investigation, it provides an insight into the methods and procedures and create a basis for interpretation of findings. As direct references on all the item are not in abundance, certain specific references along with some indirect references have been incorporated this chapter for the purpose of meaningful use.

Ditsch, D C, Trimble, R L & Wade, J M (1998) reveals that Crop production budgets are produced for conventional, reduced and no-tillage systems, and after change to reduced and no-till systems. It is concluded that although conservation tillage offers long-term advantages, in the short term yields may be reduced, forcing crop producers into financial hardship and hence the system is not used by many Kentucky farmers.

Washington, D.C., USA, National Academy Press (1998) reveals that The committee's findings point to two basic challenges: first, agricultural education must become more than vocational agriculture. Second, major revisions are needed within vocational agriculture. In working toward both goals, educators should borrow from the best current programme, while creating new ways to deliver to more students 'educational opportunities in the agricultural sciences, agribusiness, nutrition, and land resource stewardship.

Brklacich, M (1989) reveals that A sustainable food production system is defined as an agri- food system which simultaneously maintains environmental quality, provides acceptable economic and social rewards to agricultural producers, and ensures an adequate and accessible food supply. A three-dimensional framework for evaluating the sustainability of food production systems is developed comprising the resource stewardship dimension, the producer- community dimension, and the food sufficiency dimension. It was found that reductions in crop yields owing to soil erosion and climatic warming would reduce farm income potential by 26% and regional production capacity would fall short of 1981 production levels by 14%.

Grossi, R E (1993) reveals that the emergence of a sustainable agriculture will require considerable changes. A deliberate and gradual shift in policy priorities, what might be called a 'green evolution' of agriculture, toward principles of resource stewardship and marketplace economics, which must be recognized as not mutually exclusive goals, is necessary.

Fox, G (1994) reveals that the structure of property ownership does have important implications for resource stewardship and particularly, the institution of 'several property' plays an essential role in the coordination of human interaction. It is concluded, however, that further development is necessary. Clarification of the basis and contents of rights is also required.

L M; Liggett, N; Havens, D; Anderson, W C; Stark, L K (1994) reveals that the partnerships between farmers, consumers, agricultural industry representatives, environmentalists, government and university personnel in the form of focus groups can take advantage of creative group energy to jointly identify needs and set priorities for agricultural research and education. It was concluded that the focus group approach offers some novel possibilities to the land grant university system for rethinking the way in which it sets and carries out its agenda. There is a need to establish problem solving teams that include a wide range of interests, some of which may have been excluded in the past.

Colman, D (1994) reveals that neoclassical microeconomic theory of the firm and household rests on the foundations of profit maximizing and utility maximizing behaviour respectively. After exploring the issue of whether market efficiency can be improved when economic agents' behaviour is motivated by ethical objectives such as fairness and altruism, agricultural stewardship came up as an ethically motivated form of behaviour, which is profit sacrificing, but which improves economic efficiency by reducing environmental externalities.

Regan, J B (1995) reveals that the efficacy of Treflan in no till wheat, results indicated that Treflan 10 gave adequate control of S. indica with high dosage and/or appropriate timing, and gave better control when incorporated than Treflan MTF, whilst residues were maintained at acceptable levels. Best residue conservation occurred when cultivation was done by disc, combination tool, rotary hoe or Hay buster under cutter. The subsequent wheat crop was unaffected by any of the treatments.

Ticknor, W D (1996) reveals that the first Arkansas Forest Resources Distinguished Lecture deals with excellence in resource stewardship, land ethics, ecosystem management, systems thought and several other issues relevant to modern forest management in the particular context of Arkansas.

El Baltagy, A (1996) reveals that poverty, agricultural employment and migration; food consumption and production; natural resource stewardship; and global economic integration and food security are of great concern and in which we should focus to meet the challenge of the current levels of development, to halt the escalating food gap, and tackle rural poverty and natural resource management problems.

Hoogheem, T J (1997) reveals that Efforts underway within Monsanto Company to further reduce the unintended impacts to the water environment from the use of its crop protection products are described. Examples of these efforts include monitoring and other surveillance activities, investigation and remediation, new farming practices, training and prevention

Gilmore, D W(1997) reveals that the ecosystem management is a resource use concept that has evolved as a result of increasing public awareness and society's concerns regarding the future availability of goods and services from the forest resource. This paper reviews the evolution of the ecosystem management philosophy and its use as an approach toward holistic resource stewardship, particularly of forests in North America, to meet the ecological, social and economic needs of society.

Ribaudo, M; Caswell, M F (1999) reveals that the US Department of Agriculture has a long history of promoting natural resource stewardship through various forms of voluntary assistance such as education, technical assistance and cost sharing. Water quality programmes that have arisen from the Clean Water Act and state water quality protection laws generally force farmers through command and control or other incentive mechanisms to adopt alternative management practices.

Holsman, R H (2000) reveals that the expanded role of wildlife management and conservation in addressing diverse social and ecological goals necessitates an examination of the role that hunters have served and can serve in ecosystem stewardship. The notion of ecological stewardship is presented as a moral norm a personal ethic of right and wrong that influences our attitudes and behaviours toward conservation issues and one that requires consideration of the impacts of our lifestyle choices on environmental quality.

Wander, M. M., Drinkwater, L. E. (2000) reveals that Educational materials that highlight soil contributions to farm, landscape and global functioning coupled with dialogue between practitioners, scientists, and policy makers can communicate the importance of soil quality to sustainability.Successful soil quality efforts will relate soil properties to soilfunction in a way that fosters stewardship among individuals and builds public support for policies that promote soil management to ensure agriculture, industry, and the natural environment are sustained. Strategies and priorities are expected to vary according to audience, land use constraints and the intended scale of application.

Lichtenberg, E (2000) reveals that the source of environmental quality problems and the strength of incentives for resource stewardship that may incidentally induce farmers to protect environmental quality and environmental policy design under the features that are pervasive in agriculture: (1) heterogeneity caused by resource variability; and (2) uncertainty and also the effects of interactions between agricultural, environmental, and resource policies are important implications to design policy distinctive nature of environmental quality problems in agriculture, an industry based on the extraction of highly variable natural resources under stochastic conditions.

Rotterdam, Netherlands, A.A., Balkema, (2000)reveals that the problems of sustaining the quality of land that has already been reclaimed is possible through minimizing the long term liability of reclaimed land; a strategy for determining acceptable sediment yield for reclaimed mine lands; rainfall erosion on opencast coal mine lands ecological perspective; erosion control principles and some technical options; soil stewardship on reclaimed coal lands

Murombedzi, J (2001) reveals that Zimbabwe has devolved ownership of wildlife to people in the communal areas, and thereby promoted stewardship of wildlife through the production of benefits for the participating communities. In this regard natural resource management, household accumulation of CAMPFIRE revenues, and local conceptualizations of 'development' in the Masoka area were taken into account.

Carruthers, D V (2001) reveals that many indigenous art forms demonstrate characteristics associated with triumphs in agro ecology, social forestry, and other rural initiatives, i.e., community revitalization, resource stewardship, traditional ecological knowledge, identity, biological and cultural diversity, and grassroots participation. However, field research in rural Michoacan and Oaxaca presents a sobering counterpoint, finding most peasant artisans entangled, both as agents and victims, in worrisome patterns of degradation, depletion, and exploitation. Global and domestic market failures, predatory bossism, embedded inequities, lack of cooperative success, and political exclusion largely undermine in practice the ecological potential of indigenous art.

Cooke, R J; Park, J R (2001) reveals that the Countryside Stewardship Scheme offers financial incentives to farmers and landowners to improve the conservation and amenity value of English farmland. These payments often impact on 'core business' income and therefore entry decision requires careful financial appraisal. A decision support tool (the Countryside Stewardship Farm Assessor) was constructed for use by farmers considering entry to the scheme.

Delate, K (2002) reveals that systems theory holds that biological systems, such as agroecosystems, consist of integrated units of people, plants, animals, soil, insects and microorganisms, and each subsystem provides feedback to the other. The goal of each long term agro ecological research (LTAR) was to examine the short and long term physical, biological, and socioeconomic effects of organic and conventional farming systems. By establishing long term experiments, longer crop rotations, typical of organic farms, provide yield stability, improve plant protection, and enhance soil health and economic benefits compared to conventional systems with shorter rotations and greater off farm inputs kind of hypothesis was tested.

Feedback from the local farm associations that are responsible for farm stewardship and farm finances is inherent in the LTAR process.

Letey, J; Sojka, R E; Upchurch, D R; Cassel, D K; Olson, K R; Payne, W A; Petrie, S E; Price, G H; Reginato, R J; Scott, H D; Smethurst, P J; Triplett, G B (2003) reveals that the alternative view that emphasis on known principles of soil management is a better expenditure of limited resources for soil stewardship than developing and deploying subjective indices which fail to integrate across the necessary spectrum of management outcomes. Emphasis should be directed toward using available technical information to motivate and educate farmers on management practices that optimize the combined goals of high crop production, low environmental degradation, and a sustained resource.

Miller, F P (2003) reveals that the future wellbeing of humanity is inexorably linked to our understanding of the natural world and to its sustainable management. Agriculture and managed renewable natural resources account for >90% of the nation's terrestrial ecosystems. The management and stewardship of these vast ecosystems along with the ecological goods and services derived therefrom are huge and noble responsibilities. Natural resources management and ecosystem stewardship must be ecology based, including the human dimensions of ecosystem manipulation and natural resource use. Until economic systems recognize the value of environmental investments and marketability of ecological services, adoption of sustainable ecosystem management will be short circuited by rational economic decision making under current economic systems and personal values on the part of those responsible for managing these lands.

S K Acharjee, U.Ghosh, A.Biswas and A.K.Murah (2004) reveals that the role of women in promoting and supporting farm forestry, directly and indirectly, is unquestionably important and essential .Operational holding, land under farm forestry, women's participation and information use index substantively influenced the consequent variable, perceived benefits of farm forestry.

Landers, J N (2004) reveals that the zero tillage requires a holistic approach towards crop rotations, biological controls and environmental impacts, where farmer led research priorities lead to quicker and more cost effective results. The public image of the farmer as "steward of the land" is crucial to receiving environmental services payments. Farmers and their organizations can contribute considerably to research, especially if they have some effective control in deciding research priorities.

Stonehouse, D P (2004) reveals that Expanding urbanization/industrialization in southern Ontario is outcompeting agriculture for use of some of Canada's best farmland. On the remaining farmland, there has been a trend to consolidate into fewer and larger enterprises, to specialize production lines, to mechanize and automate, to increase usage of imported synthetic inputs, all at the expense of environmental protection, and natural resource stewardship and Sustainability of the agri food system should include the dimension of rural farm community viability, organic farming, reduced input farming , farmer education and extension assistance, financial assistance, cross compliance measures, and compulsion backed by litigation and penalties.

Franklin, S B; Kupfer, J A;Grubaugh, J W; Kennedy, M L (2004) reveals that A test was done to check the functional ability of ecological land types (ELTs) for biodiversity stewardship. Woody and herbaceous flora as well as her petofauna and avifauna communities had substantial differences between upland and lowland forests. However, none of the faunal communities distinguished among upland (ELTs). In addition, herbaceous taxa also failed to distinguish upland ELTs. Results suggest the present use of ELTs at Natchez Trace State Forest (NTSF) will not be a helpful guide to land stewardship focusing on biodiversity. The disturbance history of the forest and the mobility of animals are given as potential explanations for a poor relationship between abiotic ELTs and the resident biota.

Patra M K, Acharjee S K and Chakraborty S K (2005) reveals that the erosion of local fish resources has both immediate and long term impacts. Modernized agriculture and expanding urbanization have not only polluted the surroundings but also threatened the existence of key fisheries taxa.

Leaman, D J (2006) reveals that the uncounted additional species are used in the growing cosmetics and botanicals industries. The great majority of these materials is provided by collection from wild resources. Building on successes in the development of sustainable harvest and resource stewardship standards in different sectors such as timber, non-wood forest products, organic agriculture and marine aquarium fish, an initiative has recently been launched to develop an international standard and criteria for good practice in the sustainable wild collection of medicinal and aromatic plants.

Mitchell, D P; Norton, T W; Grenfell, R I; Woodgate, P W (2007) reveals that the extent to which past natural resource management (NRM) investment in Australia has actually improved the condition of natural resources is unclear. Many lessons remain to be drawn that can benefit NRM and help empower decision makers responsible for resource stewardship. Lack of information to support decision making, the small returns on agriculture in marginal areas, poor allocation of funding, reluctance to release information and a lack of whole of government approach to management were seen to be important barriers.

Davies, B B; Hodge, I D (2007) reveals that understanding how farmers conceive of their environmental rights and responsibilities is important in helping to design agri environmental measures that can complement, rather than conflict, with the underlying normative assumptions that farmers themselves hold. Through Q analysis identifies five distinct perspectives on notions of agricultural stewardship were identified termed as Environmentalists, Progressives, Commodity Conservationists, Jeffersonian and Yeomen. It was found that Q methodology can be a valuable way of demonstrating the nature of the mental frameworks of actors in a particular context and this enables us to formulate some important questions regarding the motivations of land managers in the context of a rapidly changing rural policy.

Campbell, A (2008) reveals that soil stewardship offers a crucial key to unlock the apparent tradeoffs between food security, energy security, water security and managing runaway climate change.Improvingsoil stewardship is fundamental to the world being able to feed itself while improving land, water and energy productivity, and reducing greenhouse gas emissions.

Overloop, P J van; Miltenburg, I J; Clemmens, A J; Strand, R J (2008) reveal that in the Western part of the United States of America, more and more irrigation canal networks are being modernized, which often includes some form of real time automatic water level control. The objective of the controller is to maintain the water levels in the canal pools at a given target level. The control loop is repeated at a fixed time interval. To attenuate noise and undesired signal contents, filtering of the water level measurements is required. The reasons that irrigation districts decide to automate their canals include: improved water delivery service, improved efficiency of water delivery, improved utilization of labor, and water resource stewardship.

Lowthe Thomas, S C; Clarke, A; Clapham, S J; Morgan, M; Semere, T; Slater, F M; Smith, R (2008) reveal that energy crops are becoming increasingly important for biomass production and farmland diversification and, in Wales, are being considering for integration into agricultural stewardship schemes under an ongoing review of land management schemes. Trials suggested that reed canary grass crops are economically competitive with other upland land uses, have comparable yields to other energy grasses and have alternative non energy markets. The biodiversity of the crop and potential of the crop for environmental protection are discussed in the context of growing the crop as a low input system.

Grigg, N S (2008) reveals that coordination across geographic areas is recognized as a special challenge of integration and emerging efforts toward integration in some countries will bear close scrutiny to measure their success. Regardless of the ultimate success in promoting the concept of integrated water resources management, it communicates to stakeholders the notion that water decisions require balance and the promotion of shared values, such as conservation and the alleviation of poverty. If institutional barriers and confusion about the concept are overcome, actual improvements in water stewardship may result.

Fayram, A H; Schenborn, D A; Hennessy, J M; Nate, N A; Schmalz, P J (2009) revealed that resource stewardship requires solid fisheries science and long term goals. Agency responsiveness requires a willingness to work with local interests on both long and short term needs within the framework of what is biologically possible. Finally, agency accountability for results includes the obligation to quantitatively monitor, evaluate, and communicate the consequences of management decisions to stakeholders and decision makers.

Ahnstrom, J; Hockert, J; Bergea, H L; Francis, C A; Skelton, P; Hallgren, L (2009) reveal that farmers operate in a tight financial situation, and in parts of the world they are highly dependent on government subsidies, and cannot afford to risk losing that support. Use of conservation practices is most effective when these are understood in the context of the individual farm, and decisions are rooted in land and resource stewardship and long-term concerns about health of the farm and the soil. As new agri environmental schemes are planned, agricultural development specialists need to recognize the complexity of farmer attitudes, the importance of location and individual farmer circumstances, and the multiple factors that influence decisions.

Lai PoHsin; Sorice, M G; Nepal, S K; Cheng ChiaKuen (2009) reveal that National Park Service (NPS), is associated with the task with the mission to provide open access for quality outdoor recreation and maintain the ecological integrity of the park system. In addition to management practices of education and restrictions, building a sense of natural resource stewardship among visitors facilitated the NPS ability to react to this task.

Lertzman, K (2009) reveals that reframing the concept from management to management system provides a tool for better understanding how social and ecological dynamics act as coupled drivers of managed ecosystems. Seen from this perspective, there are strong parallels between the traditional resource management systems of indigenous peoples and western science based management systems. Stewardship is a western concept which resonates with the foundations of traditional resource management systems. Both systems of management can be understood on gradients of human influence on ecosystems.

10

Methods and Approaches of Study

The deliberation on the methodology has been made to understand to concept, methods, and techniques which utilized to design the study, collection of information, analysis of the data and interpretation of the findings for the revelation of truths and formulation of theories. This present chapter deals with the method and a procedure used in the study and consists of eight main parts.

Locale of research	Sampling design
Pilot study	Preparation of interview schedule
Pre-testing of schedule	The technique of field data collection
Attributes and their measurement	Statistical tools used for data analysis

10.1 Locale of Research

Two blocks namely Dinhata-I, and Dinhata-II of the Cooch behar district in West Bengal were selected for the study. The area had been selected for the study because of

A. The researcher's close familiarity with respect to area, people officials and local dialects

B. The ample opportunity to generate relevant data due to the close proximity of the area with many research institutions and state Agricultural Universities;

C. The highly cooperative, responsive respondents;

D. The profuse scope to get relevant information about the changes that took place in and around the tea garden ecosystem in past decades;

E. Experienced, well versed and venturesome respondents

10.2 Sampling Design

The purposive, as well as simple random sampling techniques, were adopted for the present study. It may be termed as a multistage random sampling procedure. The districts, blocks, and villages were purposively selected for the study. The district Cooch behar and the Blocks Dinhata-I, and Dinhata-IIwere considered. An exhaustive list of respondents was prepared with the help of block officials for villages. From the prepared list to twenty respondents were selected randomly from each block for final data collection.

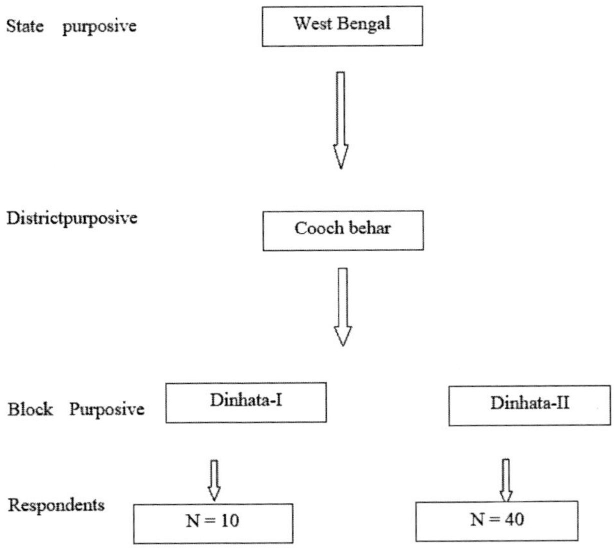

Diagram: The schematic diagram (Multistage Random Sampling) features the sampling scheme of the State, District, Block, Villages, and Respondents.

10.3 Pilot Study

A pilot study was conducted in the selected blocks before constructing the data collecting devices. In the course of this survey informal discussion was carried out with some farmers, local leaders and extension agents of the localities. An outline of the socio-economic background of the farmers, community members, stakeholders of the concerned areas, their opinion towards changing agricultural practices from convention to conservation both in terms of economy and social ecology.

10.4 Preparation of Interview Schedule

On the basis of findings of the pilot study a preliminary interview schedule was formed with the help of literature, discussion by the assistance of Chairman of Advisory Committee. The interview schedule consisted of two major parts according to the specific objectives of the study.

10.5 Pre-testing of Schedule

Before starting final data collection, the entire schedule was pretested for elimination, addition, and alternation with non-sample respondents of the study area.

10.6 The Technique of Field Data Collection

The total three blocks were personally interviewed during puja vacation. The items were asked in Bengali as well as in Hindi version in simple term so that the members could understand easily. The entries were done in the schedule by student investigator himself at the time of interview.

10.7 Attributes and their Measurement

After reviewing various literature related to the field of study and consultation with the respected chairman of the Advisory Committee and other experts, a list of variables was prepared. On the basis of selected variables, a schedule was formed.

10.7.1 Independent variables

1. Age (X_1): Chronological age has been considered for the study.
2. Education (X_2): Education status of the respondents has been considered for the study and denoted by the real numbers (i.e. 1, 2, 3...etc.)
3. Family Education (X_3): Average family education status of the respondents has been considered for the study and denoted by the real numbers (i.e. 1, 2, 3... etc.)
4. Years in CA(X_4): Years dedicated by the respondent under CA has been considered for the study.
5. Homestead land(X_5): Total area of the lands under homestead, barren and pond area has been considered for the study and denoted by the real numbers (i.e. 1, 2, 3...etc.).
6. Agricultural land (X_6): Total agricultural land (under CA as well as conventional) has been considered for the study.
7. Land under zero tillage(X_7): The changing percentage of the area of land under zero tillage has been considered for the study.
8. Count of livestock(X_8): The decadal mean value of count of livestock has been considered for the study.
9. Average volume of cow dung(X_9): Average volume of cow dung obtained from the livestock has been considered for the study.
10. Volume of manure cow dung applied in farm land(X_{10}): The Volume of manure cow dung applied in farm land has been considered for the study.
11. Land under vegetables(X_{11}): The Land under vegetables has been considered for the study.
12. No of irrigation(X_{12}): The No of irrigation is very crucial in studying conservation practices and has been considered for the study.
13. Cropping intensity(X_{13}): The decadal mean value of cropping intensity has been considered for the study.
14. No of tillage(X_{14}): The decadal mean value of No of tillage has been considered for the study.

15. CA crops including maize(X_{15}): The decadal mean value of CA crops including maize has been considered for the study.
16. Energy consumption(X_{16}): Energy consumption has been considered for the study.
17. Count of labour: Male/Female(X_{17}): The decadal mean value of Count of labour, though they may be male or female, has been considered for the study.
18. Number of inputs use($X1_8$): The decadal mean value of Number of inputs use has been considered for the study.

10.7.2 Dependent variables

1. Yield (Y_1): The decadal mean yield has been considered for the study.
2. cost of cultivation (Y_2): The decadal mean value of cost of cultivation has been considered for the study.
3. income from prominent crop (Y_3): The decadal mean value of the income generated from the farmlands has been considered for the study.
4. B:C Ratio(Y_4): B:C Ratio have been considered for the study.
5. Business interaction (Y_5): Farmer's business interaction has been taken in ten-point scale for the study.
6. Risk bearing attitude (Y_6): Farmer's risk bearing attitude has been taken in ten-point scale for the study.
7. Investment attitude(Y_7): Farmer's investment attitude has been taken in ten-point scale for the study.
8. Profit making attitude(Y_8): Farmer's profit-making attitude of the respondents has been taken in ten-point scale for the study.
9. Enhancement of area under CA(Y_9): Enhancement of area under CA has been considered for the study.

10.8 Statistical Analysis and Interpretation of Data

The role of statistics in research is to function as a tool in designing research, analysing its data and drawing conclusions of their form. Most research studies result in a large volume of raw data, which must be suitably reduced so that the same can be read easily and can be used for further analysis. Clearly, the science of statistics cannot be ignored by any research worker, even though he may not have occasion to use the statistical method in all their details and ramifications.

After the collection of data, data were processed and analysed in accordance with the outline laid down for the purpose at the time of developing the research plan. Processing implies editing, coding, classification, and tabulation of collected data. The main statistical techniques and tools used in the present study were-

Mean, Standard deviation, Coefficient of variation, Coefficient of correlation, Regression, Multiple regressions (Step-wise regression and backward regression), Path analysis, and Factor analysis, Principal Component Analysis and Canonical constant analysis. Maximum and Minimum

In statistics, the sample maximum and sample minimum also called the largest observation and smallest observation, are the values of the greatest and least elements of a sample. They are basic summary statistics, used in descriptive statistics such as the five-number summary and Bowley›s seven-figure summary and the associated box plot.

The minimum and the maximum value are the first and last order statistics(often denoted $X_{(1)}$ and $X_{(n)}$respectively, for a sample size of n).

If the sample hasoutliers, they necessarily include the sample maximum or sample minimum, or both, depending on whether they are extremely high or low. However, the sample maximum and minimum need not be outliers, if they are not unusually far from other observations.

Range

In statistics, the range of a set of data is the difference between the largest and smallest values. However, in descriptive statistics, this concept of the range has a more complex meaning. The range is the size of the smallest interval (statistics) which contains all the data and provides an indication of statistical dispersion. It is measured in the same units as the data. Since it only depends on two of the observations, it is most useful in representing the dispersion of small data sets.

Mean

The measure of central tendency (or statistical averages) tells us the point about which items have a tendency to cluster. Such a measure is considered as the most representative figure for the entire mass of data. The measure of central tendency is also known as a statistical average. Mean, median and mode are the most popular averages. Mean, also known as arithmetic average, is the most common measure of central tendency and may be defined as the value, which we get by dividing the total of the of various given items in a series by the total number of items, we can work it out as under:

$$\text{Mean } (\overline{X}) = \Sigma \frac{xi}{n} \text{ or}$$

Where, $\overline{X}$ = the symbol we use for the mean (pronounced as X bar)

Σ = Symbol for summation

Xi = Value of the ith item X, I = 1, 2,…….,n

N= total number of items

Mean is the simplest measurement of central tendency and is a widely used measure. Its chief use consists of summarizing the essential features of a series and in enabling data to be compared. It is amenable to algebraic treatment and is used in further statistical calculations. It is a relatively stable measure of central tendency. But it suffers from some limitations viz., it is unduly affected by extreme; it may not coincide with the actual value of an item in a series, and it may lead to strong impressions, particularly when the item values are not given with the average. However, the mean is better than other averages, especially in economic and social studies where direct quantitative measurements are possible.

Standard deviation

Standard deviation is the most widely used measure of the desperation of a series and is commonly denoted by the symbol 'σ' (pronounced as sigma). Standard deviation is defined as the square root of the average of squares of deviations when such deviation for the values of individuals' items in a series is obtained from the arithmetic average. It is worked out as under

Standard deviation (σ) = $\sqrt{\sum (X_i - \overline{X})^2 / n}$

Variance

In probability theory and statistics, the variance is the expectation of the squared deviation of a random variable from its mean. Informally, it measures how far a set of (random) numbers are spread out from their average value. Variance has a central role in statistics, where some ideas that use it include descriptive statistics, statistical inference, hypothesis testing, and goodness of fit and Monte Carlo sampling. A variance is an important tool in the sciences, where statistical analysis of data is common. The variance is the square of the standard deviation, the second central moment of a distribution, and the covariance of the random variable with itself, and it is often represented by σ^2, s^2 and Var(X).

Coefficient of variation

When we divide the standard deviation by the arithmetic average of the series, the resulting quantity is known as the coefficient of standard deviation, which happens to be relative measure, and is often used for comparing with a similar measure of other series. When this coefficient of standard deviation is multiplied by 100, the resulting figure is known as the coefficient of variation. Sometimes, we work out the square of standard deviation, known as variance, which is frequently used in the context of analysis of variance.

The standard deviation (along with several related measures like variance, coefficient of variation, etc.) is used mostly in research studies and is regarded as a very satisfactory measure of dispersion in a series. It is amenable to mathematical manipulation because the algebraic signs are not ignored in its calculation (as we ignore in case of mean deviation). It is less affected by fluctuations of sampling.

These advantages made standard deviation and its coefficient a very popular measure of the scatteredness of a series. It is popularly used in the context of estimation and testing of hypotheses.

Coefficient of correlation

So far we have dealt with those statistical measures that we use in the context of the univariate population i.e., the population consisting of a measure of only one variable. In the case of bivariate or multivariate populations, we often wish to know the relation of the two and/or more variables in the data to one another.

Kurt Pearson's coefficient of correlation (or simple correlation) is the most widely used method of measuring the degree of relationship between two variables. This coefficient assumes the following:

i. That there is a linear relationship between the two variables.

ii. That the two variables are causally related which means that one of the variables is independent and the other one is dependent; and

iii. A large number of independent causes are operating in both variables so as to produce a normal distribution.

Karl Pearson's coefficient of correlation

$$r = \sum(\text{Xi} - \overline{X})(\text{Yi} - \overline{Y}) / \text{n}.\sigma\text{x}.\sigma\text{y}$$

Where, Xi = i^{th} value of X variable

$\overline{X}$ = mean of X

Yi = i^{th} value of Y variable

$\overline{Y}$ = mean of Y

n = number of pairs of observations of X and Y

σx = Standard deviation of X

σ**y** = Standard deviation of Y

Karl Pearson's coefficient of correlation is also known as the product moment correlation coefficient. The value of 'r' lies between ±l. Positive values of r indicate a positive correlation between the two variables (i.e., changes in both variables take place in the same direction), whereas negative values of 'r' indicate negative correlation i.e. changes in the two variables taking place in the opposite directions. A zero value of r indicates that there is no association between the two variables When r (+) 1, it indicates perfect positive correlation and when it is (-1), it indicates perfect negative correlation, meaning thereby that variations in independent variable (X) explain 100% of the variations in the dependent variable (Y), we can also say that for a unit change in the independent variable, if there happens to be a constant change in the dependent variable in the same direction, the correlation will be termed as perfect positive. But if such change occurs in the opposite direction, the correlation will be termed as perfect negative. The value of 'r' nearer to +1 or -1 indicates a high degree of correlation between the two variables.

Regression

The correlation coefficient only expresses association and by itself tells us nothing about the causal relationships of the variates. Thus, purely from the knowledge that two variates x and y are correlated, we cannot say whether variation in x is the cause or the results of the variation in y or whether the association results from mutual dependence of the two variates or from common causes affecting both of them.

Similarly, the more existence of a high value of correlation coefficient is not necessarily indicative of an underlying relationship between the two varieties.

The underlying relation between y and x in a bivariate population can be expressed in the form of a mathematical equation known as regression equation and is said to represent the regression of the variate y and the variate x (**Panse and Sukhatme,** 1967).

If y is the dependent variable and x is the independent variable, then the linear regression equation can be written as

$$y = a + bx$$

The values of a and b can be obtained by the method of least squares which consists of minimizing the expression.

$\Sigma(y1\text{-}1\text{-}bxl)^2$ with respect to a and b

The values of a and b

$$a = y\text{-}bx$$

$$b = \frac{\sum xy - (\sum xy)(\sum yi)}{\sum xi^2 - \frac{(\sum xi)^2}{n}}$$

The regression line can now be written as

$$Y = y - bx + bx + bx \text{ or } y - y = b\ (x\text{-}x)$$

Where b is the regression coefficient.

Multiple regression analysis

When there are two or more than two independent variables, the analysis concerning relationship is known as multiple correlations and the equation describing such a relationship as the multiple regression equation. We here explain multiple correlations and regression taking only two independent variables and one dependent variable (convenient computer programmes exist for dealing with a great number of variables). In this situation the results are interpreted as shown below:

Multiple regression equation assumes the form

$$\hat{Y} = a + b_1X_1 + b_2X_2$$

Where X_1 and X_2 are two independent variables and Y being the dependent variable, and the constants a, b_1 and b_2 can be solved by solving the following three normal equations:

$$\Sigma Yi = na + b_1\Sigma X_1i + b_2\Sigma X_2i$$

$$\Sigma X_1i\ Yi = a\ \Sigma X_1i + b_1\Sigma X^2_1i + b_2\Sigma X_1i\ X_2i$$

$$\Sigma X_2i\ Y_2i = a\ \Sigma X_2i + b_1\Sigma X_1i\ X_2i + b_2\Sigma X_2i^2$$

(It may be noted that the number of normal equations would depend upon the number of independent variables. If there are 2 independent variables, then 3 equations, if there are 3 independent variables then 4 equations and so on, are used).

In multiple regression analysis, the regression coefficients (viz., b_1, b_2) become less reliable as the degree of correlation between the independent variables (viz., X_1, X_2) increases. If there is a high degree of correlation between independent variables, we have a problem with what is commonly described as the problem of multicollinearity. In such a situation we should use only one set of the independent variable to make our estimate. In fact, adding a second variable, say X_2, that is correlated with the first variable, say X_1, distorts the value of the regression coefficients. Nevertheless, the prediction for the dependent variable can be made even when multi-collinearity is present, but in such a situation enough care should be taken in selecting the independent variables to estimate a dependent variable so as to ensure that multicollinearity is reduced to the minimum.

With more than one independent variable, we may make a difference between the collective effect of the two independent variables and the individual effect of each of them taken separately. The collective effect is given by the coefficient of multiple correlations.

In multiple regressions, we form a linear composite of explanatory variables in such a way that it has a maximum correlation with a criterion variable. The main objective of using this technique is to predict the variability of the dependent variable based on its covariance with all the independent variables. One can predict the level of the dependent phenomenon through multiple regression analysis models, given the levels of independent variables.

Stepwise multiple regression

Stepwise regression is a variation of multiple regression which provides a means of choosing independent variables that yield the best prediction possible with the fewest independent variables. It permits the user to solve a sequence of one or more multiple linear regression problems by stepwise application of the least square method. At each step in the analysis, a variable is added or removed which results in the greatest production in the error sum of squares (**Burroughs Corporation,** 1975)

According to **Draper and Smith** (1981), the method of stepwise multiple the analysis is to insert variables in turn until the regression equation is satisfactory. The order of insertion is determined by using the partial correlation coefficient as a measure of the importance of variables not yet in the equation. The programme, according to **Burroughs Corporation** (1975), first forms a correlation matrix, finds the best predictor (the independent variable having the highest correlation with the criterion variable) and performs a regression analysis with this predictor. Then. the second best predictor (independent variable having the second highest correlation with the criterion) is found and regression analysis using the multiple correlations of the two best predictors is performed, and so on. At any given step, the group of predictors being used is not necessarily the best group of that size (i.e. the particular group of independent variables does not necessarily have the highest multiple correlations with the criterion that any group of this size does).

Rather, this group contains the variables that have the highest individual correlation with the criterion. Significance of a variable that is being considered for entrance into the regression equation is measured by the F-statistic. If F is too small (less than F 'include'), the variable is not added to the regression equation. Include statement establishes the minimum value of the F-statistic required for the inclusion of a variable in the regression equation. In the example which follows, the F-value for inclusion was 0.01.

Significance of variables already in the regression equation may change as new variables are entered. This significance of the variables currently in the equation is also measured by the F-statistic. If F is too small (less than F 'delete'), the variable is not added to the equation. Delete establishes the value of the F statistic below which the variable is deleted from the regression equation. Here the F-value for deletion was 0.005. The 'tolerance' level specified is used as control of degeneracy. Degeneracy, when a variable entered into the equation, is a linear combination of variables present in the equation. Tolerance statement establishes the maximum value a pivoted element may attain while still allowing its associated variable to be brought into equation. A variable is not brought into the regression equation if its associated pivoted element is below the specified tolerance level, which was 0.001 in the present example.

Factor analysis

Factor analysis is by far the most often used multivariate technique of research studies, especially pertaining to social and behavioural sciences. It is a technique applicable when there is a systematic interdependence in finding out something more fundamental or latent which creates this commonality. For instance, we might have data, say, about an individual's income, education, occupation, and dwelling area and want to infer from these some factor (such as social class), which summarizes the commonality of all the said four variables. The techniques used for such purpose is generally described as factor analysis. Factor analysis, thus, seeks to resolve a large set of measured variables in terms of relatively few categories, known as factors. This technique allows the researcher to group variables into factor (based on the correlation between variables) and the factor so derived may be treated as new variables (often termed as latent variables) and their value derived by summing the values of the original variables which have been grouped into the factor. The meaning and name of such new variable are subjectively determined by the researcher. Since the factors happen to be combinations of data, the coordinates of each observation or variable are measured to obtain what are called factor loadings. Such factor loadings represent the correlation between the particular variables and the factor and are usually placed in a matrix of correlations between the variable and the factors.

Canonical Correlation Analysis

Canonical correlation analysis is used to identify and measure the associations among two sets of variables. Canonical correlation is appropriate in the same situations where multiple regression would be, but where are there are multiple inter-correlated outcome variables. Canonical correlation analysis determines a set of canonical variates, orthogonal linear combinations of the variables within each set that best explain the variability both within and between sets. Canonical correlation analysis can be defined as the problem of finding two sets of basis vectors, one for x and the other for y, such that the correlations between the projections of the variables onto these basis vectors are mutually maximized. Multiple regression analysis is a multivariate technique which can predict the value of a single (metric) dependent variable from a linear function of a set of independent variables. For some research problems, however, interest may not centre on a single dependent variable; rather, the researcher may be interested in relationships between sets of multiple dependent and multiple independent variables. **Canonical correlation analysis** is a multivariate statistical model that facilitates the study of interrelationships among sets of multiple dependent variables and multiple independent variables. Whereas multiple regression predicts a single dependent variable from a set of multiple independent variables, canonical correlation simultaneously predicts multiple dependent variables from multiple independent variables. Canonical correlation places the fewest restrictions on the types of data on which it operates. Because the other techniques impose more rigid restrictions, it is generally believed that the information obtained from them is of higher quality and may be presented in a more interpretable manner. For this reason, many researchers view canonical correlation as a last-ditch effort, to be used when all other higher-level techniques have been exhausted. But in situations with multiple dependent and independent variables, canonical correlation is the most appropriate and powerful multivariate technique. It has gain acceptance in many fields and represents a useful tool for multivariate analysis, particularly as interest has spread to considering multiple dependent variables.

To clarify further the nature of canonical correlation, let us consider an extension of a simple example of multiple regression analysis. Assume that a survey was conducted to understand the relationships between family size and income as predictors of the number of credit cards a family would hold. Such a problem would involve examining the relationship between two independent variables and a single dependent variable.

Suppose the researcher then became interested in the broader concept of credit usage. To measure credit usage, the researcher considered not only the number of credit cards held by the family but also the family's average monthly dollar charges on all credit cards. These two measures were felt to give a much better perspective on a family's credit card usage. Readers interested in the approach of using multiple indicators to represent a concept are referred to discussions of Factor Analysis and Structural Equation Modelling The problem now involves predicting two dependent measures simultaneously (number of credit cards and average dollar charges). Multiple regressions is capable of handling only a single dependent variable. Multivariate analysis of variance could be used, but only if all of the independent variables were non-metric, which is not the case in this problem. Canonical correlation represents the only technique available for examining the relationship with multiple dependent variables. By using canonical correlation analysis, the researcher creates a composite measure of credit usage that consists of both dependent variables, rather than having to compute a separate regression equation for each of the dependent variables. The result of applying canonical correlation is a measure of the strength of the relationship between two sets of multiple variables (canonical variates). The measure of the strength of the relationship between the two variates is expressed as a canonical correlation coefficient (Rc). The researcher now has two results of interest: the canonical variates representing the optimal linear combinations of dependent and independent variables; and the canonical correlation representing the Canonical correlation analysis is the most generalized member of the family of multivariate statistical techniques. It is directly related to several dependence methods. Similar to regression, canonical correlation's goal is to quantify the strength of the relationship, in this case between the two sets of variables (independent and dependent). It corresponds to factor analysis in the creation of composites of variables. It also resembles discriminant analysis in its ability to determine independent dimensions (similar to discriminant functions) for each variable set, in this situation with the objective of producing the maximum correlation between the dimensions. Thus, canonical correlation identifies the optimum structure or dimensionality of each variable set that maximizes the relationship between independent and dependent variable sets. Canonical correlation analysis deals with the association between the composites of sets of multiple dependent and independent variables. In doing so, it develops a number of independent **canonical functions** that maximize the correlation between the **linear composites,** also known as **canonical variates,** which are sets of dependent and independent variables. Each canonical function is actually based on the correlation between two canonical variates, one variate for the dependent variables and one for the independent variables. Another unique feature of canonical correlation is that the variates are derived to maximize their correlation. Moreover, canonical relationship between them. Correlation does not stop with the derivation of a single relationship between the sets of variables. Instead, a number of canonical functions (pairs of canonical variates) may be derived.

Cluster Analysis

Cluster analysis is a class of techniques that are used to classify objects or cases into relative groups called clusters. Cluster analysis is also called classification analysis or numerical taxonomy. In cluster analysis, there is no prior information about the group or cluster membership for any of the objects.

Cluster Analysis has been used in marketing for various purposes. Segmentation of consumers in cluster analysis is used on the basis of benefits sought from the purchase of the product. It can be used to identify homogeneous groups of buyers.

Cluster analysis involves formulating a problem, selecting a distance measure, selecting a clustering procedure, deciding the number of clusters, interpreting the profile clusters and finally, assessing the validity of clustering.

The variables on which the cluster analysis is to be done should be selected by keeping past research in mind. It should also be selected by theory, the hypotheses being tested, and the judgment of the researcher. An appropriate measure of distance or similarity should be selected; the most commonly used measure is the Euclidean distance or its square.

Clustering procedures in cluster analysis may be hierarchical, non-hierarchical, or a two-step procedure. A hierarchical procedure in cluster analysis is characterized by the development of a tree-like structure. A hierarchical procedure can be agglomerative or divisive. Agglomerative methods in cluster analysis consist of linkage methods, variance methods, and centroid methods. Linkage methods in cluster analysis are comprised of single linkage, complete linkage, and average linkage.

The non-hierarchical methods in cluster analysis are frequently referred to as K means clustering. The two-step procedure can automatically determine the optimal number of clusters by comparing the values of model choice criteria across different clustering solutions. The choice of clustering procedure and the choice of distance measure are interrelated. The relative sizes of clusters in cluster analysis should be meaningful. The clusters should be interpreted in terms of cluster centroids.

Path Analysis:

The objective of doing Path Analysis is to get a clear picture of the direct and indirect effects of the independent variables on the dependent variable. Variables, through which substantial indirect effects are channelled, are also found out.

Singh and Chaudhary (1977), defined path coefficient as the ratio of the standard deviation of the effect due to a given cause to the total standard deviation of the effect i.e. if Y is the effect and x1 is the cause, the path coefficient for the path from cause x1 to the effect Y is $\sigma x1/\sigma y$

It is advisable to do path analysis with only those variables which have significant effects on the dependent variable. This may be done by restoring to multiple regression analysis and selecting those independent variables whose partial b value are significant. This shall enhance the clarity of the path analysis.

11

Case Study and Hard Evidences

Table 1: Descriptive Statistics of Independent Variable with Respect to Minimum, Maximum, Mean, Standard Deviation of Values, Variance, and Coefficient of Variance:

Variable	Minimum	Maximum	Mean	Std. Deviation	Variance	CV (%)
Age	21.000	69.000	40.760	12.484	155.860	30.629
Family size	1.000	8.000	4.500	1.502	2.255	33.371
Education	3.000	12.000	7.660	2.592	6.719	33.839
Years in CA	1.000	8.000	2.420	1.386	1.922	57.288
Homestead land	0.000	20.000	4.226	3.716	13.807	87.926
Agricultural land	0.500	19.000	4.476	3.709	13.759	82.872
Land under zero tillage	0.000	7.000	1.641	1.707	2.914	104.029
Count of livestock	0.000	8.000	1.920	1.605	2.575	83.579
Average volume of cow dung	0.000	22.000	6.200	6.707	44.980	108.172
Volume of manure cow dung applied in farmland	0.000	9.000	3.580	2.598	6.749	72.564
Land under vegetables	0.000	6.000	1.354	1.408	1.983	104.009
Change in no of irrigation	-1.000	-0.250	-0.623	0.205	0.042	-32.945
Cropping intensity	0.000	400.000	104.438	78.837	6215.219	75.486
Change in no of tillage	-1.000	0.333	-0.900	0.310	0.096	-34.493
Change in no of CA crop including maize	1.000	9.000	3.640	1.893	3.582	51.995

Variable	Minimum	Maximum	Mean	Std. Deviation	Variance	CV (%)
Energy consumption (fuel wood/ firewood)	40.000	171.200	100.012	35.980	1294.527	35.975
Change in count of labour: male/ female	-1.000	-0.250	-0.674	0.249	0.062	-36.985
Change in number of inputs use	0.140	1.000	0.499	0.222	0.049	44.466

Table 1 presents the descriptive statistics of independent variables to provide the visualization of distribution and nature of the variables of the respondents of the study for the given research location. These statistics are main for highlighting the basic characters of the variable in terms of minimum, maximum, mean, standard deviation, variance and CV value. CV value indicates the consistency of the variable in terms of is distributive character. The lesser is the value of the CV, the higher would be the consistency in this distribution. Again lower is the CV value of the variables, the better would be the compliance in terms of the normal distribution curve. Then it will be more suitable for parametric test.

Here the mean **Age** of the variable (x_1) as the table depicts 40.76 years, means most of the respondents are of mid age group and the CV value is 30.629 per cent, means it is presenting a higher level of consistency.

The distribution pattern of variable, **Family size** (x_2), depicts the maximum family size of respondents is 8 and the minimum is 1. The mean size of family is 4.5 with S.D of 1.502. The CV value is 33.371 per cent which indicates the distribution of variables is highly consistent.

The distribution pattern of variable, **Education** (x_3), depicts the maximum Education of the respondents is 12 and the minimum is 3. The mean Education is 7.660 with S.D of 2.592. The CV value is 33.839 per cent which indicates the distribution of variables is highly consistent.

The distribution pattern of variable, **Years in CA** (x_4), depicts the maximum Years in CA of respondents is 8 and the minimum is 1. The mean Years in CA is 2.42 with S.D of 1.386. The CV value is 57.288 per cent which indicates the distribution of variables is fairly consistent.

The distribution pattern of variable, **Homestead land**(x_5), depicts the maximum Homestead land of respondents is 20 and the minimum is 0. The mean Homestead landis 4.226 with S.D of 3.716. The CV value is 87.926 per cent which indicates the distribution of variables is fairly consistent.

The distribution pattern of variable, **Agricultural land** (x_6), depicts the maximum Agricultural land of respondents is 19 and the minimum is 0.5. The mean Agricultural land is 4.476 with S.D of 3.709. The CV value is 82.872 per cent which indicates the distribution of variables is fairly consistent.

The distribution pattern of variable, **Land under zero tillage** (x_7), depicts the maximum Land under zero tillage of respondents is 7 and the minimum is 0. The mean Land under zero tillage is 1,641 with S.D of 1.707. The CV value is 104.029 per cent which indicates the distribution of variables is fairly consistent.

The distribution pattern of variable, **Count of livestock** (x_8), depicts the maximum Count of livestock of respondents is 8 and the minimum is 0. The mean Count of livestock is 1.92 with S.D of 1.605. The CV value is 83.579 per cent which indicates the distribution of variables is fairly consistent.

The distribution pattern of variable, **Average volume of cow dung** (x_9), depicts the maximum Average volume of cow dung of respondents is 22 and the minimum is 0. The mean Average volume of cow dung is 6.2 with S.D of 6.707. The CV value is 108.172 per cent which indicates the distribution of variables is fairly consistent.

The distribution pattern of variable, **Volume of cow dung manure applied in farmland** (x_{10}), depicts the maximum Volume of cow dung manure applied in farm of respondents is 9 and the minimum is 0. The mean Volume of cow dung manure applied in farm is 3.58 with S.D of 2.598. The CV value is 72.564 per cent which indicates the distribution of variables is fairly consistent.

The distribution pattern of variable, **Land under vegetable** (x_{11}), depicts the maximum Land under vegetable of respondents is 6 and the minimum is 0. The mean Land under vegetable is 1.354 with S.D of 1.408. The CV value is 104.009 per cent which indicates the distribution of variables is fairly consistent.

The distribution pattern of variable, **Change in number of irrigation** (x_{12}), depicts the maximum Change in number of irrigation of respondents is -0.25 and the minimum is -1. The mean Change in number of irrigation is -0.623 with S.D of 0.205. The CV value is -32.945 per cent which indicates the distribution of variables is highly consistent.

The distribution pattern of variable, **Cropping intensity** (x_{13}), depicts the maximum Cropping intensity of respondents is 400 and the minimum is 0. The mean Cropping intensity is 104.438 with S.D of 78.837. The CV value is 75.486 per cent which indicates the distribution of variables is fairly consistent.

The distribution pattern of variable, **Change in number of tillage** (x_{14}), depicts the maximum Change in number of tillage of respondents is 0.333 and the minimum is -1.0. The mean Change in number of tillage is -0.9 with S.D of 0.31. The CV value is -34.493 per cent which indicates the distribution of variables is highly consistent.

The distribution pattern of variable, **Change in number of CA crop including maize** (x_{15}), depicts the maximum Change in number of CA crop like maize of respondents is 9 and the minimum is 1. The mean Change in number of CA crop like maize is 3.64 with S.D of 1.893. The CV value is 51.995 per cent which indicates the distribution of variables is highly consistent.

The distribution pattern of variable, **Energy consumption** (fuel wood/ fire wood) ($x1_6$), depicts the maximum Energy consumption (fuel wood/ fire wood) of respondents is 171.2 and the minimum is 40. The mean Energy consumption (fuel wood/ fire wood) is 100.012 with S.D of 35.98. The CV value is 35.975 per cent which indicates the distribution of variables is highly consistent.

The distribution pattern of variable, **Change in count of labour** (x_{17}), depicts the maximum Change in count of labour by the respondents is -0.25 and the minimum is -1.0. The mean Change in count of labour is -0.674 with S.D of 0.249. The CV value is -36.985 per cent which indicates the distribution of variables is highly consistent.

Tabel 2: Coefficient of Correlation (r): Yield(Y_1) and 18 independent variables

Sl. No.	Independent Variables	'r' Value	Remarks
1	Age	-0.104	
2	Family size	-0.029	
3	Education	0.077	
4	Years in CA	0.502	**
5	Homestead land	-0.010	
6	Agricultural land	0.039	
7	Land under zero tillage	0.048	
8	Count of livestock	0.220	
9	Average volume of cow dung	-0.005	
10	Volume of manure cow dung applied in farm land	0.376	**
11	Land under vegetables	0.154	
12	Change in no of irrigation	-0.017	
13	Cropping intensity	-0.057	
14	Change in no of tillage	-0.181	
15	Change in no of CA crop like maize	-0.117	
16	Energy consumption fuel wood/fire wood)	0.026	
17	Change in count of labour: male/female	0.021	
18	Change in number of inputs use	0.048	

**Correlation is significant at the 0.01 level

*Correlation is significant at the 0.05 level

The distribution pattern of variable, **Change in number of inputs use** (x_{18}), depicts the maximum Change in number of inputs use by the respondents is 1 and the minimum is 0.14. The mean Change in number of inputs use is 0.499 with S.D of 0.222. The CV value is 44.466 per cent which indicates the distribution of variables is highly consistent.

Table 2 presents the coefficient of correlation between yield (decadal mean yield) (Y_1) and 18 independent variables. It depicts that two variables, Years in CA (X_4) and Volume of cow dung manure applied (X_{10}) have recorded positive and significant correlation with the dependent variable, decadal mean yield.

Revelation:

The result suggests that whenever the respondents are having more years in CA, their average yield in their CA farm has increased. The higher is the experience in CA, the better would be the ecological interpretation of CA by the farmers and at the same time it would be reflected the sustainable growth in their CA farm. The variable, Volume of cow dung manure applied has been reflected in the positive change of yield in CA. The cow dung manure has brought unique properties of augmentation of organic carbon level in soil, retention of soil moisture, improvement in electrical conductivity and at the same time it has the source of natural nitrogen to the crop. Whenever the nutrient management of the CA farm has been supported by organic manure pro Tactile period, it improves the ecological resilience and services of the soil under cultivation.

Table 3: Multiple Regression Analysis: Yield (Y_1) Vs. 18 Causal Variables

Sl. No.	Variables	Reg. Coef. B	S.E. B	Beta	t Value
1	Age	0.000	0.005	-0.005	-0.033
2	Family size	0.024	0.045	0.094	0.521
3	Education	0.032	0.022	0.218	1.424
4	Years in CA	0.174	0.044	0.637	3.993
5	Homestead land	-0.022	0.035	-0.215	-0.620
6	Agricultural land	0.016	0.025	0.159	0.644
7	Land under zero tillage	-0.012	0.045	-0.053	-0.263
8	Count of livestock	0.043	0.037	0.182	1.175
9	Average volume of cow dung	-0.015	0.011	-0.258	-1.358
10	Volume of manure cow dung applied in farmland	0.056	0.022	0.381	2.497
11	Land under vegetables	-0.012	0.054	-0.044	-0.220
12	Change in no of irrigation	0.150	0.293	0.081	0.511
13	Cropping intensity	0.000	0.001	-0.049	-0.208
14	Change in no of tillage	0.168	0.207	0.137	0.811

Sl. No.	Variables	Reg. Coef. B	S.E. B	Beta	t Value
15	Change in no of CA crop like maize	-0.029	0.035	-0.146	-0.829
16	Energy consumption (fuel wood/fire wood)	-0.001	0.002	-0.088	-0.504
17	Change in count of labour	-0.012	0.258	-0.008	-0.047
18	Change in number of inputs use	-0.092	0.247	-0.054	-0.374

R square: 50.60%

The standard error of the estimate: 0.335

Table 3 presents the Multiple Regression Analysis (pool model) : Yield (decadal mean yield) (Y_1) vs 18 causal variable(X_1 to X_{18}).

Revelation:

The results reveal that the co efficient of correlation value, at displayed in the table 2, have been well established and indicated by the beta values of the respective variable in the table 3. So this variable has also exerted substantive functional effect on consequent variable decadal mean yield (Y_1).

The R square value being 50.60 per cent, it can be inferred that these 18 causal variables together had contributed to 50.60 per cent variance in the consequent variable, decadal mean yield (Y_1).

Table 4: Stepwise Regression Analysis: Yield (Y_1) Vs. 18 Causal Variables

Sl. No	Variables	Reg.coef. B	S.E. B	Beta	t value
1	Years in CA	0.130	0.032	0.476	4.089
2	Volume of manure cow dung applied in farm land(quintal/bigha)	0.050	0.017	0.340	2.923

R square: 36.70%

The standard error of the estimate: 0.308

Table 4 presents the stepwise regression analysis wherein the variable having highest substantive contribution to Y1 have been retained at the last step and these two variables contributed 36.70 per cent variance embedded in Y1. So it can be concluded that these two variables together have contributed ((36.70/50.60) *100) per cent, means 72.53 per cent of the total variance as explained by all the 18 variables together as reflected in the pool model of regression analysis. These two variables offer tremendous strategic implication for the North Bengal agriculture. The application of cow dung manure has a tremendous role in the retention of soil and harvest quality and at the same time it will be helpful for the restoration of ecological resilience. Whenever the yield has been sustainable, it indicates that the CA has been well socialized in north Bengal agriculture and has started showing entrepreneurial dividend over time. The farmers having a deeper profile of experiences in CA can be treated as opinion leader as well as community figurehead for conservation agriculture.

Table 5: Path Analysis: Decomposition of Total Effect into Direct, Indirect and Residual Effect: Yield (Y_1)

Sl. No	Variables	Total Effect	Direct Effect	Indirect Effect	Highest Indirect Effect
1	Age	-0.104	-0.005	-0.099	-0.083(x4)
2	Family size	-0.029	0.094	-0.123	0.078(x4)
3	Education	0.077	0.218	-0.141	-0.054(x5)
4	Years in CA	0.502	0.638	-0.136	-0.068(x5)
5	Homestead land (bigha)	-0.010	-0.211	0.201	0.207(x4)
6	Agricultural land	0.039	0.157	-0.118	-0.134(x5)
7	Land under zero tillage	0.048	-0.055	0.103	0.202(x4)
8	Count of livestock	0.220	0.183	0.037	0.063(x10)
9	Average volume of cow dung(kg /day/cattle)	-0.005	-0.259	0.254	0.115(x4)
10	Volume of manure cow dung applied in farm land(quintal/ bigha)	0.376	0.381	-0.005	-0.053(x14)
11	Land under vegetables	0.154	-0.045	0.199	0.203(x4)
12	Change in no of irrigation	-0.017	0.082	-0.099	-0.084(x4)
13	Cropping intensity	-0.057	-0.052	-0.005	-0.096(x5)
14	Change in no of tillage	-0.181	0.137	-0.318	-0.147(x10)
15	Change in no of CA crop like maize	-0.117	-0.145	0.028	0.052(x4)
16	Energy consumption in kilogram/month(fuel wood/fire wood)	0.026	-0.088	0.114	0.101(x4)
17	Change in count of labour/ bigha:male/female	0.021	-0.008	0.029	-0.082(x4)
18	Change in number of inputs use	0.048	-0.054	0.102	0.062(x9)

Residual effect: 0.493

Table 5 presents the path analysis and it has decomposed the total effect (r value) into direct (beta value), indirect and residual effect.

It further reveals that the highest direct effect has been exalted by the exogenous variable, years in CA followed by volume of cow dung manure applied in farm land. So, this has re-established the functional contribution as reflected in the previous table.

Interestingly, the highest indirect effect has been routed through the variable change in the numbers of tillage. It is so interesting that whenever the numbers of tillage are reduced, the decadal mean yield has been improved (-0.318). So, it is amazingly in full compliance with the CA principle wherein the farmers of north Bengal have proven that with the reduction of numbers of tillage, the yield can be scaled up.

It has also been observed that, the highest indirect effect of as many as 9 exogenous variables have been routed through the variable Years in CA (X4). These 18 variables reaffirm, the swashbuckling effect of these solitary variables on the rest of other exogenous variables are ultimately impacting on the consequent variable, decadal mean yield (Y_1). The residual effect being 48.60 per cent, it can be inferred that even with the combination of thee 18 exogenous variables, 48.60 per cent variance embedded in Y_1 variable could not be explained.

This could suggest that inclusion of more numbers of the relevant variables with higher consistency level would have contributed higher level of explicable variance embedded with the consequent variable decadal mean yield (Y_1).

Table 6: Coefficient of Correlation (r): Cost of cultivation(y_2)

Sl. No.	Independent Variables	'r' Value	Remarks
1	Age	0.027	
2	Family size	0.165	
3	Education	0.027	
4	Years in CA	0.039	
5	Homestead land	-0.042	
6	Agricultural land	-0.048	
7	Land under zero tillage	0.050	
8	Count of livestock	-0.163	
9	Average volume of cow dung	0.029	
10	Volume of manure cow dung applied in farmland	0.080	
11	Land under vegetables	-0.025	
12	Change in no of irrigation	0.059	
13	Cropping intensity	-0.012	
14	Change in no of tillage	0.147	
15	Change in no of CA crop like maize	0.824	**
16	Energy consumption (fuel wood/ firewood)	0.097	
17	Change in count of labour: male/female	-0.554	**
18	Change in number of inputs use	-0.138	

**Correlation is significant at the 0.01 level

*Correlation is significant at the 0.05 level

Table presents the coefficient of correlation between cost of cultivation (decadal mean) (Y_2) and 18 independent variables. It depicts those two variables, Change in CA crop including maize (x_{15}) and Change in count of labour: male/female (x_{17}) have recorded positive and significant correlation with the dependent variable, cost of cultivation.

Revelation:

Every crop has got a cost implication. Maize involves a lower cost of cultivation than summer rice. Whenever summer rice is replaced by maize, the cost of cultivation is expected to reduce.

It has been found that the variable change in count of labour/bigha has recorded a significant but negative correlation with cost of cultivation. Whenever more numbers of agricultural labours are involved, it is expected that the cost of cultivation will hike. So that is this kind of changes has been noticed as a change during pandemic.

Table 7: Multiple Regression Analysis: Cost of cultivation(y_2) Vs. 18 Causal Variables

Sl. No.	Variables	Reg. Coef. B	S.E. B	Beta	t Value
1	Age	0.001	0.004	0.021	0.192
2	Family size	-0.003	0.039	-0.011	-0.088
3	Education	0.025	0.019	0.134	1.314
4	Years in CA	-0.010	0.038	-0.028	-0.261
5	Homestead land	0.018	0.031	0.138	0.597
6	Agricultural land	-0.017	0.022	-0.125	-0.759
7	Land under zero tillage	-0.051	0.039	-0.176	-1.316
8	Count of livestock	0.046	0.032	0.148	1.440
9	Average volume of cow dung	-0.001	0.009	-0.013	-0.103
10	Volume of manure cow dung applied in farm land	0.014	0.019	0.072	0.712
11	Land under vegetables	0.034	0.046	0.097	0.735
12	Change in no of irrigation	0.284	0.254	0.118	1.119
13	Cropping intensity	-0.001	0.001	-0.201	-1.296
14	Change in no of tillage	0.127	0.179	0.080	0.710
15	Change in no of CA crop like maize	0.214	0.031	0.822	6.980
16	Energy consumption (fuel wood/ fire wood)	0.001	0.002	0.099	0.857
17	Change in count of labour: male/ female	-0.306	0.224	-0.154	-1.367
18	Change in number of inputs use	-0.249	0.214	-0.112	-1.162

R square: 78.10%

The standard error of the estimate: 0.291

Table presents the Multiple Regression Analysis (pool model): cost of cultivation (decadal mean) (Y_2) vs 18 causal variables (x_1 to x_{18}).

Revelation:

The highest substantive effect in cultivation has been exerted by the variable change in no. of CA crop including maize. Already being discussed that a change of crop has got both energy and cost implication. Here we have found that area under maize has expanded in recent part in North Bengal and it has included a huge cost implication.

The R square value being 78.10 per cent, it can be inferred that these 18 causal variables together had contributed to 78.10 per cent variance in the consequent variable, cost of cultivation(Y_2).

Table 8: Stepwise Regression Analysis: Cost of cultivation(y_2) Vs. 18 Causal Variables

Sl. No	Variables	Reg.coef. B	S.E. B	Beta	t value
1	Change in no of CA crop like maize	0.215	0.021	0.824	10.092

R square: 68.00%

The standard error of the estimate: 0.282

Change in number of CA crops including maize has been retained at the last stage of stepwise regression analysis, it has already been justified by the above interpretation. So it can be considered the most important determinant in the cost of cultivation of CA crop.

Table 9: Path Analysis: Decomposition of Total Effect into Direct, Indirect and Residual Effect: Cost of cultivation(y_2)

Sl. No	Variables	Total Effect	Direct Effect	Indirect Effect	Highest Indirect Effect
1	Age	0.027	0.021	0.006	0.041(x7)
2	Family size	0.165	-0.010	0.175	0.165(x15)
3	Education	0.027	0.135	-0.108	-0.048(x15)
4	Years in CA	0.039	-0.026	0.065	0.067(x15)
5	Homestead land	-0.042	0.136	-0.178	-0.107(x7)
6	Agricultural land	-0.048	-0.123	0.075	-0.108(x7)
7	Land under zero tillage	0.050	-0.176	0.226	0.163(x15)
8	Count of livestock	-0.163	0.149	-0.312	-0.226(x15)
9	Average volume of cow dung	0.029	-0.013	0.042	0.080(x15)
10	Volume of manure cow dung applied in farm land	0.080	0.072	0.008	0.078(x15)
11	Land under vegetables	-0.025	0.096	-0.121	-0.079(x15)

Sl. No	Variables	Total Effect	Direct Effect	Indirect Effect	Highest Indirect Effect
12	Change in no of irrigation	0.059	0.119	-0.060	-0.038(x15)
13	Cropping intensity	-0.012	-0.201	0.189	0.062(x5)
14	Change in no of tillage	0.147	0.081	0.066	0.099(x15)
15	Change in no of CA crop including maize	0.824	0.821	0.003	0.086(x17)
16	Energy consumption(fuel wood/ fire wood)	0.097	0.099	-0.002	-0.025(x18)
17	Change in count of labour:male/ female	-0.554	-0.154	-0.400	-0.456(x15)
18	Change in number of inputs use	-0.138	-0.113	-0.025	-0.067(x15)

Residual effect: 0.219

The same variable, change in the number of CA crops including maize has contributed the highest direct effect and it has already been justified. It is quite logical that when number of crops and types of crops are changing, the count of human labour will also be changing. That is how it has exerted the highest indirect effect. The variable x_{15}, the implication of which has already been interpreted, has routed the highest indirect effect as many as 11 variables have been routed through the same variable, change in number of CA crop including maize. That is how change in cropping pattern in North Bengal has got substantive impact on the cost of cultivation as a whole. The residual effect being 21.90 per cent, it can be inferred that even with the combination of thee 18 exogenous variables, 21.90 per cent variance embedded in Y_2 variable could not be explained.

Table 10: Coefficient of Correlation (r): Income from prominent crop(y_3)

Sl. No.	Independent Variables	'r' Value	Remarks
1	Age	-0.126	
2	Family size	0.001	
3	Education	0.020	
4	Years in CA	0.342	*
5	Homestead land	-0.008	
6	Agricultural land	0.038	
7	Land under zero tillage	0.009	
8	Count of livestock	0.278	
9	Average volume of cow dung	-0.082	
10	Volume of manure cow dung applied in farmland	0.243	
11	Land under vegetables	0.184	
12	Change in no of irrigation	-0.003	
13	Cropping intensity	-0.058	

Sl. No.	Independent Variables	'r' Value	Remarks
14	Change in no of tillage	-0.231	
15	Change in no of CA crop like maize	-0.299	*
16	Energy consumption (fuel wood/ firewood)	0.145	
17	Change in count of labour: male/female	0.122	
18	Change in number of inputs use	0.129	

**Correlation is significant at the 0.01 level

*Correlation is significant at the 0.05 level

Table presents the coefficient of correlation between Income from prominent crop (Y_3) and 18 independent variables. It depicts that two variables, Years in CA (x_4) and Change in number of CA crop including maize (X_{15}) have recorded positive and significant correlation with the dependent variable, income from prominent crop.

Revelation:

It has been found that with the progress in conservation agriculture in area and intensity, the income from maize has been substantially improved.so it proves that the income from maize has been directly be inclusion of maize in North Bengal.

Table 11: Multiple Regression Analysis: Income from prominent crop(y_3) Vs. 18 Causal Variables

Sl. No.	Variables	Reg. Coef. B	S.E. B	Beta	t Value
1	Age	-0.005	0.008	-0.105	-0.573
2	Family size	0.019	0.074	0.052	0.257
3	Education	0.013	0.036	0.060	0.346
4	Years in CA	0.146	0.071	0.372	2.070
5	Homestead land	-0.011	0.057	-0.075	-0.191
6	Agricultural land	0.014	0.041	0.093	0.335
7	Land under zero tillage	-0.025	0.072	-0.077	-0.340
8	Count of livestock	0.057	0.059	0.169	0.969
9	Average volume of cow dung	-0.014	0.017	-0.177	-0.827
10	Volume of manure cow dung applied in farm land	0.057	0.036	0.272	1.585
11	Land under vegetables	0.017	0.087	0.045	0.199
12	Change in no of irrigation	0.074	0.474	0.028	0.156
13	Cropping intensity	-0.001	0.002	-0.102	-0.388
14	Change in no of tillage	0.092	0.334	0.052	0.275
15	Change in no of CA crop like maize	-0.079	0.057	-0.274	-1.381

Sl. No.	Variables	Reg. Coef. B	S.E. B	Beta	t Value
16	Energy consumption (fuel wood/ firewood)	0.001	0.003	0.087	0.442
17	Change in count of labour: male/female	0.011	0.417	0.005	0.027
18	Change in number of inputs use	0.094	0.400	0.038	0.235

R square: 37.50%

The standard error of the estimate: 0.542

Table presents the Multiple Regression Analysis (pool model): Income from prominent crop (decadal mean) (Y_3) vs 18 causal variable (x_1 to x_{18}).

Revelation:

The CA farm farms having more years of experiences, has got a substantive impact on change in income. That is how the farmers in North Bengal, following CA enterprises, they are more interested in maize. Nevertheless, there has been a steady and performing market link between the maize growers and the Farmers Producers Organization (FPO). That is how the farmers are getting a steady income from maize since 10 years.

The R square value being 37.50 per cent, it can be inferred that these 18 causal variables together had contributed to 37.50 per cent variance in the consequent variable, income from prominent crops(Y_3).

Table 12: Stepwise Regression Analysis: Income from prominent crop(y_3) Vs. 18 Causal Variables

Sl. No	Variables	Reg.coef. B	S.E. B	Beta	t value
1	Years in CA	0.145	0.051	0.369	2.864
2	Change in no of CA crop like maize	-0.095	0.037	-0.329	-2.555

R square: 22.50%

The standard error of the estimate: 0.491

Years in CA and Change in number of CA crops including maize have been retained at the last stage of stepwise regression analysis, it has already been justified by the above interpretation. So it can be considered the most important determinants in the income from prominent crops.

It is also observable that from the last 10 years the area under maize is increasing like anything. The reason being good market price, operating support from FPOs, huge bio masses are produced which can add values to the soil health, as source of fodder to the livestock, low water consumption. So all these reasons are acting favorably for the rapid socialization of maize enterprise in north Bengal.

Table 13: Path Analysis: Decomposition of Total Effect into Direct, Indirect and Residual Effect: Income from prominent crop(y_3)

Sl. No	Variables	Total Effect	Direct Effect	Indirect Effect	Highest Indirect Effect
1	Age	-0.126	-0.105	-0.021	-0.048(x4)
2	Family size	0.001	0.052	-0.051	-0.055(x15)
3	Education	0.020	0.059	-0.039	0.031(x1)
4	Years in CA	0.342	0.372	-0.030	-0.032(x9)
5	Homestead land	-0.008	-0.073	0.065	0.121(x4)
6	Agricultural land	0.038	0.092	-0.054	-0.069(x9)
7	Land under zero tillage	0.009	-0.078	0.087	0.117(x4)
8	Count of livestock	0.278	0.169	0.109	0.075(x15)
9	Average volume of cow dung	-0.082	-0.177	0.095	0.067(x4)
10	Volume of manure cow dung applied in farm land	0.243	0.272	-0.029	-0.030(x9)
11	Land under vegetables	0.184	0.045	0.139	0.118(x4)
12	Change in no of irrigation	-0.003	0.027	-0.030	-0.057(x10)
13	Cropping intensity	-0.058	-0.103	0.045	-0.033(x5)
14	Change in no of tillage	-0.231	0.052	-0.283	-0.105(x10)
15	Change in no of CA crop like maize	-0.299	-0.274	-0.025	-0.047(x8)
16	Energy consumption (fuel wood/ fire wood)	0.145	0.087	0.058	-0.060(x10)
17	Change in count of labour:male/ female	0.122	0.006	0.116	0.152(x15)
18	Change in number of inputs use	0.129	0.038	0.091	0.042(x9)

Residual effect: 0.625

Table presents the path analysis and it has decomposed the total effect (r value) into direct (beta value), indirect and residual effect.

The highest direct effect has been exerted by years in CA. More numbers in years, better will be the experience and market network. That are coming in favour of the farmers helping them access better income from maize. Interestingly, the highest indirect effect has been routed through the variable change in the numbers of tillage.The residual effect being 62.50 per cent, it can be inferred that even with the combination of thee 18 exogenous variables, 62.50 per cent variance embedded in Y3 variable could not be explained.

Table 14: Coefficient of Correlation (r): B:C Ration(y_4)

Sl. No.	Independent Variables	'r' Value	Remarks
1	Age	-0.098	
2	Family size	-0.079	
3	Education	-0.010	
4	Years in CA	0.266	
5	Homestead land	0.003	
6	Agricultural land	0.050	
7	Land under zero tillage	-0.029	
8	Count of livestock	0.298	*
9	Average volume of cow dung	-0.104	
10	Volume of manure cow dung applied in farmland	0.098	
11	Land under vegetables	0.203	
12	Change in no of irrigation	-0.054	
13	Cropping intensity	-0.068	
14	Change in no of tillage	-0.243	
15	Change in no of CA crop like maize	-0.669	**
16	Energy consumption (fuel wood/ firewood)	0.098	
17	Change in count of labour: male/female	0.420	**
18	Change in number of inputs use	0.158	

**Correlation is significant at the 0.01 level

*Correlation is significant at the 0.05 level

Table presents the coefficient of correlation between B:C Ratio (decadal mean) (Y_4) and 18 independent variables. It depicts that three variables, count of livestock (x_8), Change in CA crop including maize (x_{15}) and Change in count of labour (X_{17}) have recorded positive and significant correlation with the dependent variable, decadal mean yield.

Revelation:

The B:C Ratio has been positively impacted by count of livestock. The more is the count of livestock, the favorable is the B:C ratio. That is inclusion of livestock has positively contributed to the improvement of income and its sustainability. Specially, in case of CA, the role and contribution of livestock are immensely important. For certain area it has also been found that the inclusion of maize, has not contributed as favorably as it was expected. The risk involve during harvesting may be one reason. And also crop residue cleaning involves lot of expenditure as well. Through the inclusion of maize cultivation, the requirement of labour has been reduced in comparison to summer rice. It has positively contributed to the B:C Ratio.

Table 15: Multiple Regression Analysis: B:C Ration(y_4) Vs. 18 Causal Variables

Sl. No.	Variables	Reg. Coef. B	S.E. B	Beta	t Value
1	Age	-0.004	0.005	-0.110	-0.794
2	Family size	-0.002	0.044	-0.007	-0.046
3	Education	-0.008	0.022	-0.045	-0.347
4	Years in CA	0.100	0.042	0.321	2.349
5	Homestead land	-0.028	0.034	-0.242	-0.813
6	Agricultural land	0.020	0.024	0.168	0.798
7	Land under zero tillage	0.008	0.043	0.032	0.187
8	Count of livestock	0.023	0.035	0.084	0.635
9	Average volume of cow dung	-0.007	0.010	-0.116	-0.712
10	Volume of manure cow dung applied in farmland	0.017	0.022	0.103	0.794
11	Land under vegetables	0.024	0.052	0.080	0.470
12	Change in no of irrigation	-0.233	0.284	-0.111	-0.819
13	Cropping intensity	0.000	0.001	0.039	0.194
14	Change in no of tillage	0.025	0.201	0.018	0.125
15	Change in no of CA crop like maize	-0.129	0.034	-0.569	-3.769
16	Energy consumption (fuel wood/firewood)	0.001	0.002	0.070	0.472
17	Change in count of labour: male/female	0.300	0.250	0.174	1.198
18	Change in number of inputs use	0.143	0.240	0.074	0.596

R square: 63.90%

The standard error of the estimate: 0.325

Table presents the Multiple Regression Analysis (pool model): B:C Ratio (decadal mean) (Y_4) vs 18 causal variable (x_1 to x_{18}).

Revelation:

The same variable, years in CA has recorded the direct effect. The implications are already being given.

The R square value being 63.90 per cent, it can be inferred that these 18 causal variables together had contributed to 63.90 per cent variance in the consequent variable, B:C Ratio(Y_4).

Table 16: Stepwise Regression Analysis: B:C Ration(y_4) Vs. 18 Causal Variables

Sl. No	Variables	Reg.coef. B	S.E. B	Beta	t value
1	Change in no of CA crop like maize	-0.158	0.022	-0.696	-7.099
2	Years in CA	0.100	0.030	0.323	3.298

R square: 55.20%

The standard error of the estimate: 0.294

The same two variables as they are impacting the income, they have been written also while considering the B:C Ratio.

Table 17: Path Analysis: Decomposition of Total Effect into Direct, Indirect and Residual Effect: B:C Ration(y4)

Sl. No	Variables	Total Effect	Direct Effect	Indirect Effect	Highest Indirect Effect
1	Age	-0.098	-0.110	0.012	-0.042(x4)
2	Family size	-0.079	-0.009	-0.070	-0.114(x15)
3	Education	-0.010	-0.047	0.037	-0.061(x5)
4	Years in CA	0.266	0.319	-0.053	-0.078(x5)
5	Homestead land	0.003	-0.240	0.243	0.106(x6)
6	Agricultural land	0.050	0.167	-0.117	-0.153(x5)
7	Land under zero tillage	-0.029	0.032	-0.061	-0.146(x5)
8	Count of livestock	0.298	0.084	0.214	0.156(x15)
9	Average volume of cow dung	-0.104	-0.114	0.010	0.065(x6)
10	Volume of manure cow dung applied in farm land	0.098	0.104	-0.006	-0.054(x15)
11	Land under vegetables	0.203	0.082	0.121	-0.125(x5)
12	Change in no of irrigation	-0.054	-0.111	0.057	0.064(x5)
13	Cropping intensity	-0.068	0.039	-0.107	-0.109(x5)
14	Change in no of tillage	-0.243	0.017	-0.260	-0.068(x15)
15	Change in no of CA crop like maize	-0.669	-0.569	-0.100	-0.096(x17)
16	Energy consumption (fuel wood/ fire wood)	0.098	0.071	0.027	0.051(x4)
17	Change in count of labour:male/ female	0.420	0.173	0.247	0.316(x15)
18	Change in number of inputs use	0.158	0.074	0.084	0.047(x15)

Residual effect: 0.362

Table presents the path analysis and it has decomposed the total effect (r value) into direct (beta value), indirect and residual effect.

The homestead land has routed the highest indirect effect of as many as 7 times. So homestead land can be an important indicator in propagating the maize entrepreneurship. The land which is consider for maize cultivation is noting but the extension of their homestead land. That is how the homestead land has played a vital role. Besides, management of maize can better be done when it is attached to homestead land. The residual effect being 36.20 per cent, it can be inferred that even with the combination of thee 18 exogenous variables, 36.20 per cent variance embedded in Y4 variable could not be explained.

Table 18: Coefficient of Correlation (r): Business interaction(y_5)

Sl. No.	Independent Variables	'r' Value	Remarks
1	Age	-0.379	**
2	Family size	0.200	
3	Education	0.207	
4	Years in CA	0.194	
5	Homestead land	0.393	**
6	Agricultural land	0.147	
7	Land under zero tillage	0.337	*
8	Count of livestock	0.018	
9	Average volume of cow dung	0.011	
10	Volume of manure cow dung applied in farmland	-0.173	
11	Land under vegetables	0.284	*
12	Change in no of irrigation	0.160	
13	Cropping intensity	0.290	*
14	Change in no of tillage	-0.093	
15	Change in no of CA crop like maize	0.261	
16	Energy consumption (fuel wood/firewood)	-0.120	
17	Change in count of labour: male/female	-0.089	
18	Change in number of inputs use	-0.118	

**Correlation is significant at the 0.01 level

*Correlation is significant at the 0.05 level

Table presents the coefficient of correlation between Business interaction (y_5) and 18 independent variables. It depicts that five variables, Age (x_1), Homestead land (x_5), Land under zero tillage (x_7), Land under vegetables(x_{11}), and Cropping intensity (x_{13}) have recorded positive and significant correlation with the dependent variable, decadal mean yield.

Revelation:

It has been found that the younger farmers are more engaged in business interaction regarding maize enterprises. The young farmers are developed a strong profile of entrepreneurship for the inclusion of maize and other CA crop. They have also recorded a positive gesture like mulching. This is to conclude that the young farmers in North Bengal are coming up with higher exuberance and commitment for commercializing CA crop including maize. Entrepreneurial management can be natural and effective when it is congruous to homestead land. Homestead land can be an important indicator in propagating the maize entrepreneurship. The land which is consider for maize cultivation is nothing but the extension of their homestead land. That is how the homestead land has played a vital role.

More numbers of land under zero tillage means less numbers of conventional ploughing. When the numbers of tillage have reduced or rationalized, the cost has also been rationalized and the whole of the enterprise has becoming more economically resilience and sustainable.

The recent trend in North Bengal is that in CA the following vegetables are increasingly being included like potato, brinjal, leafy vegetables etc. A good count of vegetables has been included in CA design and practices which has triggered up diversified and effective business interaction. In this case also the market has been created for vegetables and FPOs are extending their effective support for better and wider business interaction.

Whenever new crops are incorporated, count of vegetables are incorporated as inter-crop, the cropping intensity is bound to be increased.

Table 19: Multiple Regression Analysis: Business interaction(y_5) Vs. 18 Causal Variables

Sl. No.	Variables	Reg. Coef. B	S.E. B	Beta	t Value
1	Age	-0.058	0.025	-0.342	-2.337
2	Family size	0.661	0.227	0.471	2.909
3	Education	0.159	0.112	0.196	1.426
4	Years in CA	0.274	0.218	0.180	1.254
5	Homestead land	0.165	0.177	0.291	0.930
6	Agricultural land	-0.048	0.126	-0.084	-0.379
7	Land under zero tillage	0.117	0.223	0.095	0.526
8	Count of livestock	0.057	0.183	0.043	0.313
9	Average volume of cow dung	-0.102	0.054	-0.325	-1.904
10	Volume of manure cow dung applied in farmland	-0.164	0.111	-0.202	-1.471
11	Land under vegetables	-0.171	0.267	-0.114	-0.638
12	Change in no of irrigation	2.843	1.465	0.277	1.941
13	Cropping intensity	0.001	0.006	0.025	0.119
14	Change in no of tillage	-0.094	1.033	-0.014	-0.091

Sl. No.	Variables	Reg. Coef. B	S.E. B	Beta	t Value
15	Change in no of CA crop like maize	0.241	0.177	0.216	1.359
16	Energy consumption (fuel wood/ firewood)	-0.021	0.009	-0.357	-2.286
17	Change in count of labour: male/ female	0.632	1.289	0.075	0.491
18	Change in number of inputs use	-1.782	1.236	-0.188	-1.442

R square: 60.00%

The standard error of the estimate: 1.676

Table presents the Multiple Regression Analysis (pool model): Business interaction (Y_5) vs 18 causal variable (x_1 to x_{18}).

Revelation:

Family size has been a strong determinant for business interaction in conservation agriculture. The family having the more numbers of family member, can be a source of family labour, higher mobility and wider business interaction.

The R square value being 60.00 per cent, it can be inferred that these 18 causal variables together had contributed to 60.00 per cent variance in the consequent variable, business interaction(Y_5).

Table 20: Stepwise Regression Analysis: Business interaction(y_5) Vs. 18 Causal Variables

Sl. No	Variables	Reg.coef. B	S.E. B	Beta	t value
1	Homestead land	0.210	0.071	0.370	2.979
2	Age	-0.065	0.021	-0.388	-3.145
3	Change in no of irrigation	2.745	1.244	0.267	2.207
4	Family size	0.374	0.172	0.267	2.177

R square: 38.80%

The standard error of the estimate: 1.720

Homestead land can be an important indicator in propagating the maize entrepreneurship and it has been already proven.

When cropping intensity is increased, it is obvious that the irrigation facilities are extended.When cropping intensity is increasing, the crop diversity is soaring up It is expected that that land under zero tillage is increasing. The classical rainfed agro ecosystem of North Bengal has substantially been changed into irrigated agro-ecosystem to respond to the socialization and entrepreneurial conservation agriculture. The maize and some other crops becoming rainfall sensitive, the requirement of irrigation is reduced.

Family size has been a strong determinant for business interaction in conservation agriculture. The family having the more numbers of family member, can be a source of family labour, higher mobility and wider business interaction

Table 21: Path Analysis: Decomposition of Total Effect into Direct, Indirect and Residual Effect: Business interaction(y_5)

Sl. No	Variables	Total Effect	Direct Effect	Indirect Effect	Highest Indirect Effect
1	Age	-0.379	-0.341	-0.038	0.122(x2)
2	Family size	0.200	0.472	-0.272	-0.155(x16)
3	Education	0.207	0.197	0.010	0.101(x1)
4	Years in CA	0.194	0.181	0.013	0.095(x5)
5	Homestead land	0.393	0.292	0.101	-0.074(x12)
6	Agricultural land	0.147	-0.084	0.231	0.185(x5)
7	Land under zero tillage	0.337	0.095	0.242	0.178(x5)
8	Count of livestock	0.018	0.043	-0.025	-0.059(x15)
9	Average volume of cow dung	0.011	-0.326	0.337	0.101(x16)
10	Volume of manure cow dung applied in farm land	-0.173	-0.201	0.028	0.079(x16)
11	Land under vegetables	0.284	-0.116	0.400	0.152(x5)
12	Change in no of irrigation	0.160	0.278	-0.118	-0.078(x5)
13	Cropping intensity	0.290	0.024	0.266	0.133(x5)
14	Change in no of tillage	-0.093	-0.015	-0.078	-0.125(x1)
15	Change in no of CA crop including maize	0.261	0.214	0.047	0.095(x2)
16	Energy consumption (fuel wood/fire wood)	-0.120	-0.359	0.239	0.204(x2)
17	Change in count of labour:male/ female	-0.089	0.074	-0.163	-0.119(x15)
18	Change in number of inputs use	-0.118	-0.188	0.070	-0.081(x16)

Residual effect: 0.399

Table presents the path analysis and it has decomposed the total effect (r value) into direct (beta value), indirect and residual effect.

The homestead land has routed the highest indirect effect of as many as 6 times. So, homestead land can be an important indicator in propagating the maize entrepreneurship.The land which is consider for maize cultivation is nothing but the extension of their homestead land. That is how the homestead land has played a vital role. Besides, management of maize can better be done when it is attached to homestead land.

The residual effect being 39.90 per cent, it can be inferred that even with the combination of thee 18 exogenous variables, 39.90 per cent variance embedded in Y5 variable could not be explained.

Table 22: Coefficient of Correlation (r): Risk bearing attitude(y_6)

Sl. No.	Independent Variables	'r' Value	Remarks
1	Age	-0.135	
2	Family size	0.033	
3	Education	0.224	
4	Years in CA	0.231	
5	Homestead land	0.393	**
6	Agricultural land	0.191	
7	Land under zero tillage	0.118	
8	Count of livestock	-0.119	
9	Average volume of cow dung	0.150	
10	Volume of manure cow dung applied in farmland	-0.104	
11	Land under vegetables	0.387	**
12	Change in no of irrigation	-0.412	**
13	Cropping intensity	0.068	
14	Change in no of tillage	-0.135	
15	Change in no of CA crop like maize	-0.083	
16	Energy consumption (fuel wood/firewood)	-0.111	
17	Change in count of labour: male/female	0.060	
18	Change in number of inputs use	-0.549	**

**Correlation is significant at the 0.01 level

*Correlation is significant at the 0.05 level

Table presents the coefficient of correlation between Risk bearing attitude(Y_6) and 18 independent variables. It depicts that four variables, Homestead land (x_5), Land under vegetables(x_{11}), Change in number of irrigation (x_{12}) and Change in number of inputs use (x_{18}) have recorded positive and significant correlation with the dependent variable, decadal mean yield.

Revelation:

Whenever the maize is grown in extended homestead land, the risk bearing attitude will be more, management will be proper and intensive. Whenever crop is managed in a contiguous homestead land,it is expected that the management will be better hence risk bearing ability will be more. During covid pandemic the cost of vegetable has grown like anything, almost 300 times. So, the possibility of pilferage is more whenever the land is away from the home. That is how management and protection of CA crop is better whenever it is being followed in a contiguous homestead land.

Vegetables ensure better return because vegetables are not a choice, but a mandatory food stuff. That is how in recent past the cost of vegetables has grown up kike anything. So, it has helped the farmer to get over the cost of cultivation and their experiencing happy return during this time.

CA has contributed to the reduction of irrigation requirement. Hence it has been favorable for water conservation and better soil health.

The basic principle of CA is that to reduce the application of chemical inputs like fertilizer and pesticide, fuel consumption like diesel and electricity, and thus it reduces the cost of cultivation as well. The cost reduction will effective whenever it is supported by market price. If market price is fluctuating, then the risk will grow up and benefits of CA entrepreneurship will be countervailed.

Table 23: Multiple Regression Analysis: Risk bearing attitude(y_6) Vs. 18 Causal Variables

Sl. No.	Variables	Reg. Coef. B	S.E. B	Beta	t Value
1	Age	-0.011	0.016	-0.096	-0.688
2	Family size	0.054	0.147	0.057	0.368
3	Education	0.058	0.072	0.106	0.809
4	Years in CA	0.157	0.141	0.152	1.113
5	Homestead land	0.105	0.114	0.273	0.916
6	Agricultural land	-0.049	0.081	-0.128	-0.606
7	Land under zero tillage	-0.151	0.144	-0.181	-1.051
8	Count of livestock	-0.058	0.118	-0.065	-0.490
9	Average volume of cow dung	0.002	0.035	0.009	0.053
10	Volume of manure cow dung applied in farmland	-0.127	0.072	-0.231	-1.770
11	Land under vegetables	0.190	0.173	0.187	1.099
12	Change in no of irrigation	-1.892	0.946	-0.272	-2.000
13	Cropping intensity	-0.002	0.004	-0.094	-0.470
14	Change in no of tillage	-0.488	0.667	-0.106	-0.732
15	Change in no of CA crop like maize	-0.058	0.114	-0.077	-0.508
16	Energy consumption (fuel wood/firewood)	-0.002	0.006	-0.057	-0.379
17	Change in count of labour: male/female	0.158	0.832	0.028	0.190
18	Change in number of inputs use	-3.085	0.798	-0.479	-3.867

R square: 63.70%

The standard error of the estimate: 1.082

Table presents the Multiple Regression Analysis (pool model): Risk bearing attitude (Y_6) vs 18 causal variable (x_1 to x_{18}).

Revelation:

Whenever crop is managed in a contiguous homestead land, it is expected that the management will be better and more number of vegetables will be incorporated, means crop diversity will be improved.

The R square value being 63.70 per cent, it can be inferred that these 18 causal variables together had contributed to 63.70 per cent variance in the consequent variable, Risk bearing attitude(Y_6).

Table 24: Stepwise Regression Analysis: Risk bearing attitude(y_6) Vs. 18 Causal Variables

Sl. No	Variables	Reg. coef. B	S.E. B	Beta	t value
1	Change in number of inputs use	-3.005	0.666	-0.467	-4.512
2	Land under vegetables	0.309	0.103	0.304	2.992
3	Change in no of irrigation	-2.522	0.726	-0.362	-3.474
4	Volume of manure cow dung applied in farmland	-0.117	0.057	-0.213	-2.060

R square: 54.50%

The standard error of the estimate: 1.006

The application of cow dung manure has a tremendous role in the retention of soil organic carbon and harvest quality and, at the same time it will be helpful for the restoration of ecological resilience. Whenever the yield has been sustainable, it indicates that the CA has been well socialized in north Bengal agriculture and has started showing entrepreneurial dividend over time. The farmers having a deeper profile of experiences in CA can be treated as opinion leader as well as community figurehead for conservation agriculture.

When cropping intensity is increased, it is obvious that the irrigation facilities are extended.When cropping intensity is increasing, the crop diversity is soaring up It is expected that that land under zero tillage is increasing. The classical rainfed agro ecosystem of North Bengal has substantially been changed into irrigated agro-ecosystem to respond to the socialization and entrepreneurial conservation agriculture. The maize and some other crops becoming rainfall sensitive, the requirement of irrigation is reduced.

Table 25: Path Analysis: Decomposition of Total Effect into Direct, Indirect and Residual Effect: Risk bearing attitude(y_6)

Sl. No	Variables	Total Effect	Direct Effect	Indirect Effect	Highest Indirect Effect
1	Age	-0.122	-0.076	-0.046	-0.047(x5)
2	Family size	0.052	0.077	-0.025	0.038(x11)
3	Education	0.223	0.119	0.104	0.071(x5)
4	Years in CA	0.228	0.151	0.077	0.090(x5)
5	Homestead land	0.397	0.275	0.122	-0.098(x7)
6	Agricultural land	0.192	-0.135	0.327	0.174(x5)
7	Land under zero tillage	0.124	-0.160	0.284	0.167(x5)
8	Count of livestock	-0.108	-0.040	-0.068	-0.043(x10)
9	Average volume of cow dung	0.144	-0.009	0.153	0.122(x18)
10	Volume of manure cow dung applied in farmland	-0.109	-0.236	0.127	0.055(x12)
11	Land under vegetables	0.394	0.180	0.214	0.143(x5)
12	Change in no of irrigation	-0.408	-0.266	-0.142	-0.092(x18)
13	Cropping intensity	0.069	-0.101	0.170	0.125(x5)
14	Change in no of tillage	-0.139	-0.106	-0.033	0.092(x10)
15	Change in no of CA crop like maize	-0.087	-0.068	-0.019	0.042(x18)
16	Energy consumption (fuel wood/firewood)	-0.100	-0.053	-0.047	-0.119(x18)
17	Change in count of labour: male/female	0.078	0.051	0.027	0.038(x15)
18	Change in number of inputs use	-0.556	-0.494	-0.062	-0.050(x12)

Residual effect: 0.357

Table presents the path analysis and it has decomposed the total effect (r value) into direct (beta value), indirect and residual effect.

The homestead land has routed the highest indirect effect of as many as 7 times. So, homestead land can be an important indicator in propagating the maize entrepreneurship. The land which is consider for maize cultivation is nothing but the extension of their homestead land. That is how the homestead land has played a vital role. Besides, management of maize can better be done when it is attached to homestead land. Here this variable shows better direct and indirect effect both.

The residual effect being 35.70 per cent, it can be inferred that even with the combination of thee 18 exogenous variables, 35.70 per cent variance embedded in Y6 variable could not be explained.

Table 26: Coefficient of Correlation (r): Investment attitude(y_7)

Sl. No.	Independent Variables	'r' Value	Remarks
1	Age	-0.093	
2	Family size	0.305	*
3	Education	0.064	
4	Years in CA	-0.041	
5	Homestead land	0.193	
6	Agricultural land	0.283	*
7	Land under zero tillage	0.241	
8	Count of livestock	0.219	
9	Average volume of cow dung	0.224	
10	Volume of manure cow dung applied in farmland	0.280	
11	Land under vegetables	0.324	*
12	Change in no of irrigation	-0.005	
13	Cropping intensity	-0.018	
14	Change in no of tillage	-0.259	
15	Change in no of CA crop like maize	-0.081	
16	Energy consumption (fuel wood/firewood)	0.025	
17	Change in count of labour: male/female	-0.045	
18	Change in number of inputs use	-0.225	

**Correlation is significant at the 0.01 level

*Correlation is significant at the 0.05 level

Table presents the coefficient of correlation between Investment attitude (Y_7) and 18 independent variables. It depicts that three variables, Family size (x_2), Agricultural land (x_6) and Land under vegetables (x_{11}) have recorded positive and significant correlation with the dependent variable, decadal mean yield.

Revelation:

The investment attitude is contributed by the family size and family character. That is how it has got an important positive influence on entrepreneurship.

Investment attitude depends on resource endowment. The higher will be the holding size, the higher would be the investment attitude. Better land size promotes and provokes to trasform conventional agriculture into an entrepreneurial agriculture and it is more so for conservation agriculture. CA offers an opportunity to enter into the organic farming market. That is how it is mostly happening in large size of holding.

Vegetables are a better investment as vegetables ensure better return because vegetables are not a choice, but a mandatory food stuff. That is how in recent past the cost of vegetables has grown up kike anything. So, it has helped the farmer to get over the cost of cultivation and their experiencing happy return during this time.

Table 27: Multiple Regression Analysis: Investment attitude(y_7) Vs. 18 Causal Variables

Sl. No.	Variables	Reg. Coef. B	S.E. B	Beta	t Value
1	Age	0.004	0.026	0.029	0.170
2	Family size	0.418	0.241	0.322	1.736
3	Education	0.019	0.117	0.025	0.159
4	Years in CA	-0.413	0.227	-0.303	-1.819
5	Homestead land	0.051	0.185	0.101	0.278
6	Agricultural land	-0.073	0.131	-0.143	-0.555
7	Land under zero tillage	0.195	0.235	0.176	0.830
8	Count of livestock	0.184	0.195	0.154	0.943
9	Average volume of cow dung	0.054	0.056	0.191	0.957
10	Volume of manure cow dung applied in farmland	0.205	0.116	0.281	1.765
11	Land under vegetables	0.418	0.279	0.311	1.499
12	Change in no of irrigation	1.111	1.527	0.120	0.728
13	Cropping intensity	-0.003	0.006	-0.110	-0.451
14	Change in no of tillage	-0.671	1.076	-0.110	-0.624
15	Change in no of CA crop like maize	-0.243	0.185	-0.243	-1.318
16	Energy consumption (fuel wood/firewood)	0.000	0.010	-0.004	-0.020
17	Change in count of labour: male/female	-1.635	1.371	-0.210	-1.193
18	Change in number of inputs use	-1.268	1.301	-0.148	-0.974

R square: 47.90%

The standard error of the estimate: 1.744

Table presents the Multiple Regression Analysis (pool model): Investment attitude (Y_7) vs 18 causal variable (x_1 to x_{18}).

Revelation:

Family size is better related with economic compulsion, diverse experience, mobility and family skill. So, the investment attitude is contributed by the family size and family character. That is how it has got an important positive influence on entrepreneurship. The family size has the direct effect.

The R square value being 47.90 per cent, it can be inferred that these 18 causal variables together had contributed to 47.90 per cent variance in the consequent variable, investment attitude(Y_7).

Table 28: Stepwise Regression Analysis: Investment attitude(y_7) Vs. 18 Causal Variables

Sl. No	Variables	Reg.coef. B	S.E. B	Beta	t value
1	Land under vegetables	0.456	0.178	0.339	2.553
2	Volume of manure cow dung applied in farm land	0.216	0.097	0.296	2.236

R square: 19.30%

The standard error of the estimate: 1.574

The application of cow dung manure has got a tremendous role in the retention of soil and harvest quality and at the same time it will be helpful for the restoration of ecological resilience. Whenever the yield has been sustainable, it indicates that the CA has been well socialized in north Bengal agriculture and has started showing entrepreneurial dividend over time. The farmers, having a deeper profile of experiences in CA can be treated as opinion leader as well as community figurehead for conservation agriculture.

Table 29: Path Analysis: Decomposition of Total Effect into Direct, Indirect and Residual Effect: Investment attitude(y_7)

Sl. No	Variables	Total Effect	Direct Effect	Indirect Effect	Highest Indirect Effect
1	Age	-0.093	0.029	-0.122	0.070(x2)
2	Family size	0.305	0.322	-0.017	0.065(x11)
3	Education	0.064	0.026	0.038	0.071(x11)
4	Years in CA	-0.041	-0.302	0.261	0.100(x11)
5	Homestead land	0.193	0.100	0.093	0.162(x11)
6	Agricultural land	0.283	-0.141	0.424	0.114(x11)
7	Land under zero tillage	0.241	0.175	0.066	0.114(x11)
8	Count of livestock	0.219	0.154	0.065	0.066(x15)
9	Average volume of cow dung	0.224	0.190	0.034	0.064(x2)
10	Volume of manure cow dung applied in farmland	0.280	0.280	0.000	0.043(x14)
11	Land under vegetables	0.324	0.310	0.014	-0.097(x4)
12	Change in no of irrigation	-0.005	0.120	-0.125	-0.057(x10)
13	Cropping intensity	-0.018	-0.110	0.092	0.045(x5)
14	Change in no of tillage	-0.259	-0.110	-0.149	-0.109(x10)

Sl. No	Variables	Total Effect	Direct Effect	Indirect Effect	Highest Indirect Effect
15	Change in no of CA crop like maize	-0.081	-0.243	0.162	0.118(x17)
16	Energy consumption (fuel wood/firewood)	0.025	-0.004	0.029	0.131(x2)
17	Change in count of labour: male/female	-0.045	-0.210	0.165	0.136(x15)
18	Change in number of inputs use	-0.225	-0.148	-0.077	-0.047(x9)

Residual effect: 0.522

Table presents the path analysis and it has decomposed the total effect (r value) into direct (beta value), indirect and residual effect.

Whenever a family is enriched with family skill and there is constant requirement of vegetable for the family consumption, farmers will be more interested in vegetable enterprises and by this time they have already established a good market network with the help of FPO to face better return and market price in a better extent.

The residual effect being 52.20 per cent, it can be inferred that even with the combination of thee 18 exogenous variables, 52.20 per cent variance embedded in Y7 variable could not be explained.

Table 30: Coefficient of Correlation (r): Profit making attitude(y_8)

Sl. No.	Independent Variables	'r' Value	Remarks
1	Age	-0.659	**
2	Family size	-0.080	
3	Education	0.617	**
4	Years in CA	0.054	
5	Homestead land	0.187	
6	Agricultural land	0.126	
7	Land under zero tillage	0.061	
8	Count of livestock	0.100	
9	Average volume of cow dung	0.231	
10	Volume of manure cow dung applied in farmland	0.014	
11	Land under vegetables	0.187	
12	Change in no of irrigation	-0.147	
13	Cropping intensity	0.065	
14	Change in no of tillage	-0.301	*

Sl. No.	Independent Variables	'r' Value	Remarks
15	Change in no of CA crop like maize	-0.062	
16	Energy consumption (fuel wood/firewood)	-0.133	
17	Change in count of labour: male/female	0.018	
18	Change in number of inputs use	-0.036	

**Correlation is significant at the 0.01 level

*Correlation is significant at the 0.05 level

Table presents the coefficient of correlation between profit making attitude (Y_8) and 18 independent variables. It depicts that three variables, age (x_1), education (x_3) and change in numbers of tillage (x_{14}) have recorded positive and significant correlation with the dependent variable, decadal mean yield.

Revelation:

Profit making attitude is more prominent in younger age group. They are young, active and aspiring. That is entrepreneurship is more prominent in among the young mind.

Education invites cosmopoliteness, risk bearing ability and entrepreneurship, that is how it is positively correlated.

Whenever the number of tillage is reduced, the return is more from per unit area. Boro rice has been replaced by maize which has reduced the number of tillage also. And the farming ecosystem as followed in CA not only reduces the number of tillage but also improve the water retaintivity at nutrient status of soil health. That is why after 10 years of follow up for conservation agriculture, there has been a lot of changes in soil health, productivity as well as Entrepreneurship in North Bengal farming.

The north Bengal agriculture is predominantly a rural enterprise, offers scope from rural entrepreneurship. There is no organized or large industry. That is why educated young mind are rushing after enterprising agriculture. To them CA, livestock, fishery and orchard are offering better and higher opportunity.

Table 31: Multiple Regression Analysis: Profit making attitude(y_8) Vs. 18 Causal Variables

Sl. No.	Variables	Reg. Coef. B	S.E. B	Beta	t Value
1	Age	-0.074	0.016	-0.573	-4.764
2	Family size	0.083	0.143	0.077	0.577
3	Education	0.281	0.071	0.450	3.983
4	Years in CA	-0.060	0.138	-0.052	-0.437
5	Homestead land	0.008	0.112	0.018	0.069
6	Agricultural land	-0.016	0.079	-0.038	-0.207

Sl. No.	Variables	Reg. Coef. B	S.E. B	Beta	t Value
7	Land under zero tillage	-0.200	0.141	-0.211	-1.422
8	Count of livestock	0.102	0.115	0.101	0.885
9	Average volume of cow dung	0.038	0.034	0.159	1.133
10	Volume of manure cow dung applied in farmland	-0.076	0.070	-0.121	-1.077
11	Land under vegetables	0.169	0.169	0.147	1.000
12	Change in no of irrigation	-1.104	0.924	-0.140	-1.196
13	Cropping intensity	-0.002	0.004	-0.091	-0.528
14	Change in no of tillage	-0.300	0.652	-0.058	-0.461
15	Change in no of CA crop like maize	0.075	0.112	0.088	0.674
16	Energy consumption (fuel wood/ firewood)	0.001	0.006	0.029	0.227
17	Change in count of labour: male/ female	0.602	0.813	0.093	0.740
18	Change in number of inputs use	0.00004	0.779	0.000	0.000

R square: 73.00%

The standard error of the estimate: 1.057

Table presents the Multiple Regression Analysis (pool model): Profit making attitude (Y_8) vs 18 causal variable (x_1 to x_{18}).

Revelation:

Variable, years in CA has the highest direct effect.

The R square value being 73.00 per cent, it can be inferred that these 18 causal variables together had contributed to 73.00 per cent variance in the consequent variable, profit making attitude(Y_8).

Table 32: Stepwise Regression Analysis: Profit making attitude(y_8) Vs. 18 Causal Variables

Sl. No	Variables	Reg.coef. B	S.E. B	Beta	t value
1	Age	-0.067	0.012	-0.521	-5.601
2	Education	0.288	0.058	0.462	4.964

R square: 62.90%

The standard error of the estimate:1.006

Age and education have been strong determinants for profit making in conservation agriculture. The family having the more numbers of family member, can be a source of family labour, higher mobility and wider business interaction

Table 33: Path Analysis: Decomposition of Total Effect into Direct, Indirect and Residual Effect: Profit making attitude(y_8)

Sl. No	Variables	Total Effect	Direct Effect	Indirect Effect	Highest Indirect Effect
1	Age	-0.658	-0.565	-0.093	-0.133(x3)
2	Family size	-0.061	0.069	-0.130	-0.123(x1)
3	Education	0.617	0.449	0.168	0.168(x1)
4	Years in CA	0.050	-0.050	0.100	0.070(x1)
5	Homestead land	0.190	0.012	0.178	-0.132(x7)
6	Agricultural land	0.127	-0.032	0.159	-0.134(x7)
7	Land under zero tillage	0.067	-0.216	0.283	0.144(x1)
8	Count of livestock	0.119	0.094	0.025	0.114(x1)
9	Average volume of cow dung	0.225	0.163	0.062	0.066(x1)
10	Volume of manure cow dung applied in farmland	0.009	-0.121	0.130	0.059(x1)
11	Land under vegetables	0.194	0.149	0.045	0.102(x3)
12	Change in no of irrigation	-0.140	-0.142	0.002	-0.033(x3)
13	Cropping intensity	0.066	-0.086	0.152	0.082(x3)
14	Change in no of tillage	-0.307	-0.058	-0.249	-0.219(x1)
15	Change in no of CA crop like maize	-0.067	0.087	-0.154	-0.048(x17)
16	Energy consumption (fuel wood/firewood)	-0.120	0.027	-0.147	-0.134(x1)
17	Change in count of labour: male/female	0.039	0.086	-0.047	-0.058(x1)
18	Change in number of inputs use	-0.042	0.003	-0.045	-0.040(x9)

Residual effect: 0.272

Table presents the path analysis and it has decomposed the total effect (r value) into direct (beta value), indirect and residual effect.

Profit making attitude is more prominent in younger age group. They are young, active and aspiring. That is entrepreneurship is more prominent in among the young mind. Age has the highest indirect effect.

The residual effect being 27.20 per cent, it can be inferred that even with the combination of thee 18 exogenous variables, 27.20 per cent variance embedded in Y8 variable could not be explained.

Table 34: Coefficient of Correlation (r): Enhancement area under CA(y9)

Sl. No.	Independent Variables	'r' Value	Remarks
1	Age	-0.037	
2	Family size	0.076	
3	Education	0.009	
4	Years in CA	0.174	
5	Homestead land	0.263	
6	Agricultural land	0.377	**
7	Land under zero tillage	0.148	
8	Count of livestock	-0.027	
9	Average volume of cow dung	0.606	**
10	Volume of manure cow dung applied in farmland	0.016	
11	Land under vegetables	-0.087	
12	Change in no of irrigation	-0.365	**
13	Cropping intensity	-0.106	
14	Change in no of tillage	0.067	
15	Change in no of CA crop like maize	0.013	
16	Energy consumption (fuel wood/firewood)	-0.496	**
17	Change in count of labour: male/female	0.013	
18	Change in number of inputs use	-0.082	

**Correlation is significant at the 0.01 level

*Correlation is significant at the 0.05 level

Table presents the coefficient of correlation between enhancement under CA (decadal mean) (Y_9) and 18 independent variables. It depicts that four variables, Agricultural land (x4) and Average volume of cow dung(x_9), Change in no of irrigation (x_{12}), and Energy consumption (x_{16}) have recorded positive and significant correlation with the dependent variable, decadal mean yield.

Revelation:

The area under CA and size of agricultural land are comensurating with each other. CA entrepreneurship has been inevitable and logical reality for those having bigger size of holding. CA could not be successful in small and fragmented holding. It needs a optimal and operational size of holding.

Organic manure is an essential requirement for CA. So, the family having higher livestock count and volume of cow dung production, they are the natural choice for conservation agriculture.

CA has reduced the requirement and intensity of the irrigation because it promotes the rain water harvesting and consumption, rather than ground water depletion.

CA has change the attitude of the farmer so that they have reduced the use of conventional fuel with the expansion of CA. CA ecosystem promotes ecological vision about energy consumption and its subsequent rationalization.

Table 35: Multiple Regression Analysis: Enhancement area under CA(y_9) Vs. 18 Causal Variables

Sl. No.	Variables	Reg. Coef. B	S.E. B	Beta	t Value
1	Age	0.023	0.025	0.122	0.927
2	Family size	0.290	0.230	0.183	1.260
3	Education	-0.002	0.113	-0.002	-0.017
4	Years in CA	0.315	0.221	0.183	1.422
5	Homestead land	0.015	0.180	0.024	0.086
6	Agricultural land	0.119	0.128	0.186	0.933
7	Land under zero tillage	-0.102	0.226	-0.073	-0.449
8	Count of livestock	0.037	0.185	0.025	0.198
9	Average volume of cow dung	0.132	0.054	0.372	2.430
10	Volume of manure cow dung applied in farmland	-0.176	0.113	-0.192	-1.559
11	Land under vegetables	-0.164	0.271	-0.097	-0.605
12	Change in no of irrigation	-2.730	1.486	-0.235	-1.838
13	Cropping intensity	0.001	0.006	0.023	0.121
14	Change in no of tillage	0.229	1.048	0.030	0.218
15	Change in no of CA crop like maize	-0.143	0.179	-0.114	-0.798
16	Energy consumption (fuel wood/ firewood)	-0.036	0.009	-0.544	-3.880
17	Change in count of labour: male/ female	-0.781	1.307	-0.082	-0.597
18	Change in number of inputs use	1.458	1.253	0.136	1.164

R square: 67.80%

The standard error of the estimate: 1.699

Table presents the Multiple Regression Analysis (pool model): Enhancement area under CA (decadal mean) (Y_9) vs 18 causal variable (x_1 to x_{18}).

Revelation:

Cow dung is an essential ingredient in CA as well as organic farming. This is the important determinant that is why it has retained till the last stage.

The R square value being 67.80 per cent, it can be inferred that these 18 causal variables together had contributed to 67.80 per cent variance in the consequent variable, enhancement area under CA(Y_9).

Table 36: Stepwise Regression Analysis: Enhancement area under CA(y_9) Vs. 23 Causal Variables

Sl. No	Variables	Reg.coef. B	S.E. B	Beta	t value
1	Average volume of cow dung	0.178	0.036	0.500	4.912
2	Energy consumption (fuel wood/ firewood)	-0.024	0.007	-0.358	-3.480
3	Change in no of irrigation	-3.292	1.159	-0.283	-2.841
4	Volume of manure cow dung applied in farmland	-0.189	0.093	-0.206	-2.038

R square: 58.00%
The standard error of the estimate: 1.612

The amount of average volume of cow dung and application of cow dung manure have tremendous role in the retention of soil and harvest quality and at the same time it will be helpful for the restoration of ecological resilience. Whenever the yield has been sustainable, it indicates that the CA has been well socialized in north Bengal agriculture and has started showing entrepreneurial dividend over time.

When cropping intensity is increased, it is obvious that the irrigation facilities are extended.When cropping intensity is increasing, the crop diversity is soaring up It is expected that that land under zero tillage is increasing. The classical rainfed agro ecosystem of North Bengal has substantially been changed into irrigated agro-ecosystem to respond to the socialization and entrepreneurial conservation agriculture. The maize and some other crops becoming rainfall sensitive, the requirement of irrigation is reduced.

Table 37: Path Analysis: Decomposition of Total Effect into Direct, Indirect and Residual Effect: Enhancement area under CA(y_9)

Sl. No	Variables	Total Effect	Direct Effect	Indirect Effect	Highest Indirect Effect
1	Age	-0.037	0.122	-0.159	-0.146(x16)
2	Family size	0.076	0.183	-0.107	-0.235(x16)
3	Education	0.009	-0.002	0.011	0.041(x9)
4	Years in CA	0.174	0.182	-0.008	-0.086(x16)
5	Homestead land	0.263	0.022	0.241	0.118(x6)
6	Agricultural land	0.377	0.186	0.191	0.146(x9)
7	Land under zero tillage	0.148	-0.072	0.220	0.115(x6)
8	Count of livestock	-0.027	0.024	-0.051	-0.032(x10)
9	Average volume of cow dung	0.606	0.373	0.233	0.153(x16)

Sl. No	Variables	Total Effect	Direct Effect	Indirect Effect	Highest Indirect Effect
10	Volume of manure cow dung applied in farm land	0.016	-0.191	0.207	0.120(x16)
11	Land under vegetables	-0.087	-0.097	0.010	-0.082(x16)
12	Change in no of irrigation	-0.365	-0.236	-0.129	-0.082(x16)
13	Cropping intensity	-0.106	0.024	-0.130	-0.056(x9)
14	Change in no of tillage	0.067	0.030	0.037	0.074(x10)
15	Change in no of CA crop like maize	0.013	-0.113	0.126	0.045(x17)
16	Energy consumption (fuel wood/fire wood)	-0.496	-0.544	0.048	-0.105(x9)
17	Change in count of labour:male/ female	0.013	-0.080	0.093	0.063(x15)
18	Change in number of inputs use	-0.082	0.136	-0.218	-0.122(x16)

Residual effect: 0.321

CA has got immense ecological dimension. It adds to ecological resilience and function. Expansion of CA in north Bengal has contributed towards the rationalization of energy consumption as well. For example, the summer rice, a heavy consumer of ground water has been replaced by maize (critically a rainfed crop), huge water as well as energy has been saved.

Model: Canonical Covariate Analysis: Interaction between Left Side Dependent and Right Side Independent Variables

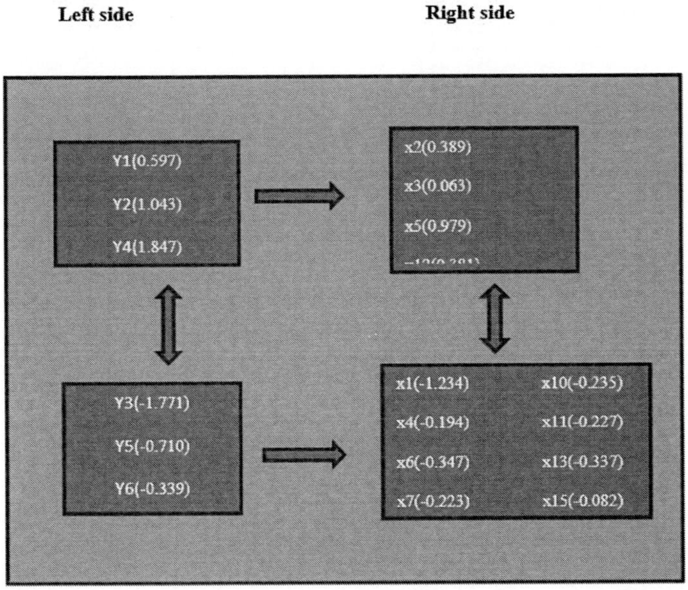

The residual effect being 32.10 per cent, it can be inferred that even with the combination of thee 18 exogenous variables, 32.10 per cent variance embedded in Y9 variable could not be explained.

This dependent variables (left side variables) Change in income from prominent crop(Y_3), Business interaction(Y_5), Risk bearing attitude(Y_6) and Profit making attitude(Y_8) have formed a sub group which reflects that these dependent variables are moving together to form a conglomeration, we can call Entrepreneurial conglomeration. This subgroup of dependent variables is being strongly affected by the following independent variables Age(x_1), Years in CA(x_4), Agricultural land(x_6), Land under zero tillage(x_7), Count of livestock(x_8), Average volume of cow dung(x_9), Volume of manure cow dung applied in farm land(x10), Land under vegetables(x11), Cropping intensity(x_{13}), Change in no of CA crop like maize(x_{15}), Energy consumption(x_{16}) and Change in number of inputs use(x_{18}).

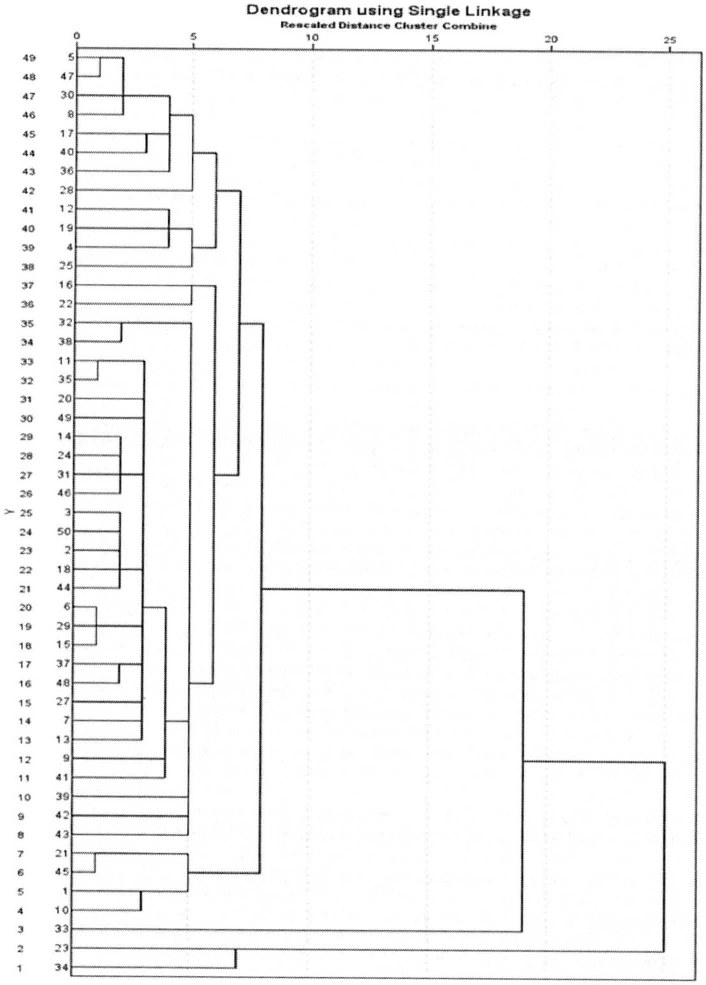

Cluster Analysis

As derived from the cluster analysis by using single linkage the whole of the respondents here has been accommodated in 5 homogenous clusters.

Cluster 1 has accommodated 5 respondents

Cluster 2 has accommodated 42 respondents

Cluster 3 has accommodated 1 respondent

Cluster 4 has accommodated 1 respondent

Cluster 5 has accommodated 1 respondent

The cluster has been formed because of the homogeneity of the respondents and distances between four clusters have been seen by measurable Euclidean distance.

So, these 5 clusters can be considered as a strategic respondent group for taking any kind of strategic intervenes.

All 50 respondents have been clustered into 5 homogeneous groups but it is important to see that cluster 2 has accommodated 42 respondents and all these respondents are homogeneous. They have formed a homogeneous group. Cluster 2 has the biggest number of respondents followed by cluster 1, has got 5 respondents and rests have 1 respondent. It has got trimendous strategic important. Basically there are 5 identifiable respondents in visual groups. Instead of 6 different respondents, now we can form 5 operating groups and the strategy for CA can be executed through this 5 different groups.

Table 38: Factor Analysis: Strategic Conglomeration of Variables into Factor

Factors	Variables	Factor loading	% of variance	Cumulative %	Factor renamed
F1	Homestead land Agricultural land Land under zero tillage Land under vegetables	0.701 0.823 0.904 0.554	16.198	16.198	CA ecosystem
F2	Change in no of CA crop like maize Change in count of labour:male/female Change in cost of cultivation Change in B:C ratio	0.916 -0.751 0.901 -0.616	13.692	29.889	Change dynamics of CA
F3	Average volume of cow dung Change in enhancement of area under CA	-0.820 -0.868	9.298	39.187	Organic driver for CA

Factors	Variables	Factor loading	% of variance	Cumulative %	Factor renamed
F4	Age Education Profit making attitude	0.681 -0.792 -0.937	7.945	47.132	Entrepreneurial motivation
F5	Years in CA Change in yield Change in income from prominent crop	-0.685 -0.927 -0.867	7.608	54.741	Social ecology of CA
F6	Count of livestock Volume of manure cow dung applied in farm land Change in no of tillage Investment attitude	0.604 0.736 -0.711 0.512	5.892	60.633	Livestock dynamics
F7	Change in number of inputs use Risk bearing attitude	0.894 -0.739	5.439	66.072	Risk bearing ability
F8	Family size Energy consumption (fuel wood/fire wood)	0.873 0.706	4.588	70.660	Family energetics
F9	Cropping intensity	0.921	4.439	75.099	
F10	Change in no of irrigation Business interaction	-0.786 -0.550	4.180	79.279	Irrigation economy

Factor 1 has accommodated the following Variables-Homestead land, Agricultural land, Land under zero tillage, Land under vegetables. This factor has contributed 16.198 per cent variance in the total interaction. This factor is renamed as **CA ecosystem.**

Factor 1 has accommodated the following variables-Change in no of CA crop like maize, Change in count of labour:male/female, Change in cost of cultivation, Change in B:C ratio. This factor has contributed 13.692 per cent variance in the total interaction. This factor is renamed as **Change dynamics of CA.**

Factor 1 has accommodated the following Variables-Average volume of cow dung, Change in enhancement of area under CA. This factor has contributed 9.298per cent variance in the total interaction. This factor is renamed as **Organic driver for CA.**

Factor 1 has accommodated the following Variables-Age, Education, Profit making attitude. This factor has contributed 7.945per cent variance in the total interaction. This factor is renamed as **Entrepreneurial motivation.**

Factor 1 has accommodated the following Variables-Years in CA, Change in yield, Change in income from prominent crop. This factor has contributed 7.608per cent variance in the total interaction. This factor is renamed as **Social ecology in CA.**

Factor 1 has accommodated the following Variables-Count of livestock, Volume of manure cow dung applied in farm land, Change in no of tillage, Investment attitude. This factor has contributed 5.892per cent variance in the total interaction. This factor is renamed as **Livestock dynamic.**

Factor 1 has accommodated the following variables-Change in number of inputs use, Risk bearing attitude. This factor has contributed 5.439per cent variance in the total interaction. This factor is renamed as **Risk bearing ability.**

Factor 1 has accommodated the following variables-Family size, Energy consumption (fuel wood/fire wood). This factor has contributed 4.588per cent variance in the total interaction. This factor is renamed as **Family energetics.**

Factor 10 has accommodated the following variables-Change in no of irrigation, Business interaction. This factor has contributed 4.180per cent variance in the total interaction. This factor is renamed as **Irrigation economy.**

These 10 factors are immensely important for strategic management of variable and their proper execution. So far as the area under research is in concern these factors are expected to play a vital role in making strategy, aggregating CA programme and describing policy implication in a micro sociological prospect.

Artificial Neural Networking

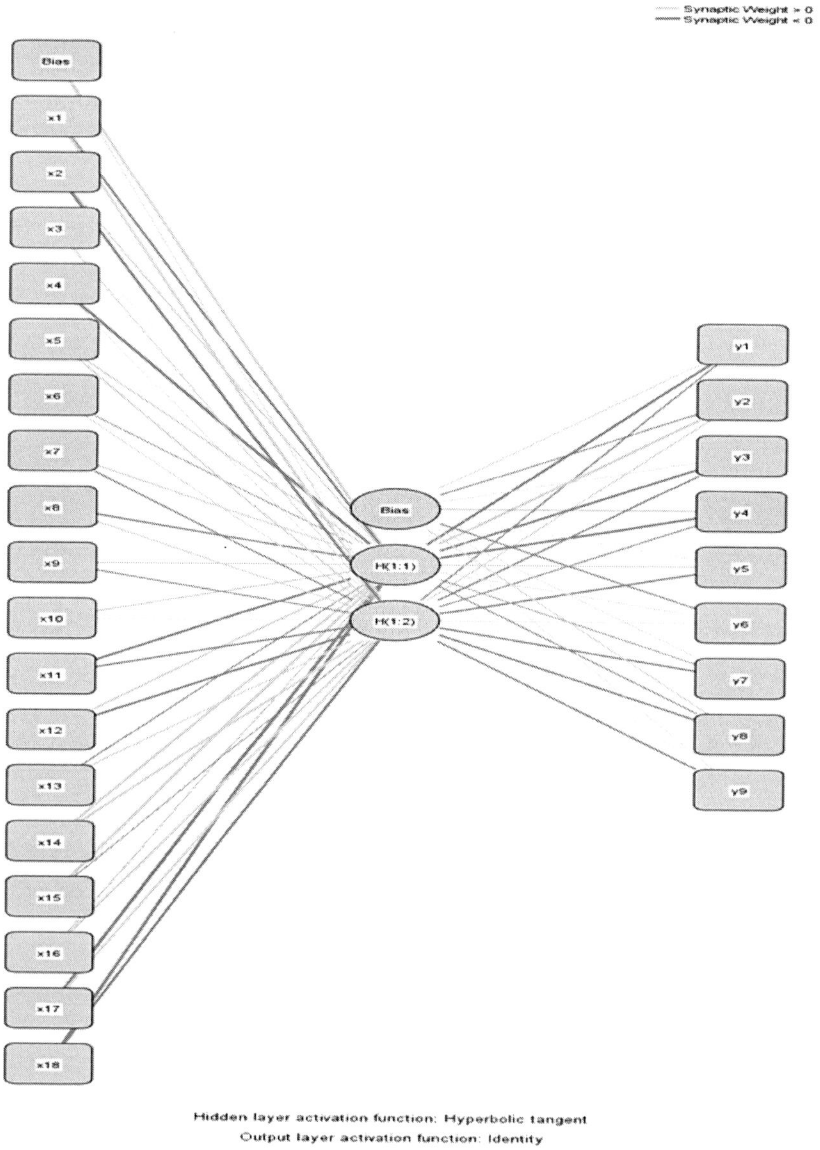

In neural network analysis, there are three layers. In the network of interaction, the following variables are exerting stronger operating impact on the right side operating variable.

Land under vegetables (X_{11}) and Number of inputs use($X1_8$), these variables have exerted (as depicted by the blue and bolder line) precise effect on another 3 output variables, y_1, y_3 and y_4. We can predict that these variables have got substantive and dominant effect (as depicted by blue and bold lines) passing through respective hidden layers to the output variables y_1, y_3 and y_4.

CA being an upcoming enterprise for north Bengal, it has been observed that inputs are properly managed, effectively used and subsequently rationalized.

So we can conclude that the type, number, cost and rationalization of input are exerting the highest effect on yield. Land under vegetables which basically have been instrumental in socialization and entrepreneurial socialization of CA. Because vegetables earn better income and it needs less water means less ground water depletion, it effectively uses the sunlight and energy. That is how it contributes to both social ecologies by providing income and biophysical ecology by providing resilience. It aims to enhance the interconnection between individual smallholder production systems and integrates production areas, forests, and ecological corridors, supporting agroecosystems (Fayad et al., 2016; Fig. 1).

Model Summary			
Training	Sum of Squares Error		134.790
	Average Overall Relative Error		0.832
	Relative Error for Scale Dependents	Change in yield	0.898
		Change in cost of cultivation	0.543
		Change in income from prominent crop no(Rs/bigha)	0.736
		Change in B:C ratio	0.461
		Business interaction	0.959
		Risk bearing attitude	1.033
		Investment attitude	0.947
		Profit making attitude	0.901
		Change in enhancement of area under CA	1.009
	Stopping Rule Used		1 consecutive step(s) with no decrease in errora
	Training Time		0:00:00.01

Testing	Sum of Squares Error		33.787
	Average Overall Relative Error		0.894
	Relative Error for Scale Dependents	Change in yield	1.332
		Change in cost of cultivation	0.698
		Change in income from prominent crop	0.879
		Change in B:C ratio	0.482
		Business interaction	0.668
		Risk bearing attitude	1.004
		Investment attitude	1.395
		Profit making attitude	1.040
		Change in enhancement of area under CA	1.308
Error computations are based on the testing sample.			

12

Summary and Conclusion

Wider adoption of CA technologies requires concerted effort of all the stakeholders in the expanded partnership and participatory approaches in which farmers' experiments and provides rapid feed-back. This would need to be supported by institutional changes that promote knowledge-sharing, flexibility and decentralized decision-making for rapid adoption of technologies to maintain production and productivity, increased food security and livelihood of the farmers. Available evidence shows continuously declining cereal crop yields during the past decade despite the large areas that are planted each year and this can be attributed to the major constraints of low and erratic rainfall, inherently low soil nutrient status, timely non-availability of inputs and lack of appropriate technologies. It is therefore important to note that intensification of crop production systems should aim at increasing crop productivity per unit area through building entrepreneurship and resilient farming systems and the use of technologies that address moisture management issues, increase the efficiency with which both external and natural recourses are used, while maintaining and improving soil fertility. However, a shift to CA has become a necessity in view of widespread problems of resource degradation, which accompanied the past strategies to enhance production with little concern for resource integrity. Farming systems should enhance resilience in order to be able to cope with change. Resilience can be viewed upon from different levels. The soil as the central regulatory center provides a valuable starting point from which to develop a broader understanding of resilience on farm level. Not only production, but many other ecological functions are dependent on soil health. Biodiversity at different levels is supporting resilience in soils, crops and animals. Rethinking farming systems in terms of resilience, without a priori rejecting solutions on ideological grounds, can overcome differences in approaches to sustainable agriculture, and bridge the gap between highly productive agriculture and conservation goals. Conservation agriculture offers an opportunity for arresting and reversing the downward spiral of resource degradation, decreasing cultivation costs and making agriculture more resource – use efficient, competitive and sustainable:

- **Sustainable intensification:** The smallholder farming sector is key to producing the food requirements of an increasing, and increasingly urban, population. Increased production must be accompanied by natural resource conservation if mankind is to have a future on this planet. CA requires specific entrepreneurship and mechanization inputs. Smallholder farmers are not often in a position to invest in expensive farm machinery and the best vehicle to provide them with the required services is via well trained and well equiped private sector entrepreneurial CA service providers.

- ❑ **Local manufacture:** Local manufacture of CA equipment is a desirable goal as it not only helps to stimulate the local economy, but also provides the opportunity to adapt technologies to local conditions be they crops, soils, climate, production systems, technical knowledge, manufacturing skills or material supply, amongst other factors.
- ❑ **Policy guidelines:** Governments will often need guidance on how to provide the best environment for nurturing a local CA equipment manufacturing industry. Financial support to the smallholder farming sector to assist farmers in the purchase of CA equipment will directly stimulate the local supply chain. The correction of anomalies affecting local manufacture, such as the existence of import duty on machine components and raw materials, but not on imported agricultural machinery, will need to be removed to provide an equitable environment for local industry.
- ❑ **Demand creation:** Efforts at creating demand for CA should be on-going. Although the public sector has a major role to play (for example in organizing field days and improving extension efforts), the private sector should also be encouraged through demonstration plots, out-grower technical support, machinery fairs and the formation and consolidation of CA farmer mutual support groups.
- ❑ **Service provision:** Given the problems of affordability and availability of CA equipment, including power sources, a promising solution is to equip and train entrepreneurial CA service providers. Equipment can be centrally procured and delivered on credit to would-be service providers. Loan payments are repaid from the resulting business activity. E-voucher systems can be employed to stimulate initial demand for the services but they should be withdrawn as soon as demand is sufficiently high.
- ❑ **Training:** Service providers not only need good CA equipment, but will also often need training in the technical aspects of its correct use, calibration and maintenance, as well as training on the managerial skills of identifying and running a successful service provision model.
- ❑ CA addresses the problem of low and erratic rainfall through the use of technologies that reduce water losses from runoff and soil evaporation and increase infiltration and soil moisture holding capacity, and improve low soil nutrient status by increasing soil carbon and nitrogen through the use of organic soil cover and legumes in rotations and interactions.

13

Future Scope of the Study

The present study has well germinated the needs of future researches and these are:

- Relegating climate change with stewardship function.
- Methodological innovations for the furtherance of advance researches.
- The socio ecological viability and efficacy of community gene bank, water auditing, disease pest scouting, scope of ICT intervention in stewardship, the ecological proficiency of traditional practices, redeeming and reengineering and soil water and biodiversity management can offer a new scope of future researches.
- The policy issues and paradigming of research interventions can go hand in hand to offer a new goner of research in extension science.
- The eminent research agenda under the aegis of stewardship and resource monitoring can incorporate energy metabolism, farm function entropy, frame working of micro ecology, geo spatial dynamics of social ecology, stress and apocalyptic impact of urbanization the magnitude and dimensions, gender and ethnic issues of forest ecology, community mobilization, auditing, empowerment and entitlement can also offer a huge future scope of extension researches.

Once our good Earth has woombed the whole of the civilization, the embryonic human society has been up scaled to a humongous global civilization but by this time our greed not the need has destroyed an inflicted both the anatomy and physiology of mother earth. Time is just ticking out and shall we remain thoughtless and mindless?

Recommendation

- Conservation education should be incorporated in the present course and curriculum in order to generate basic and advance knowledge.
- The farm innovation like community gene bank, people's biodiversity registrar, water Panchayat, folk gene consortia, soil conservation task force have to be created with necessary legal as well as statutory provision.

- Both up scaling and backstopping of technologies are essential for making entrepreneurial function a sustainable proposition and practices.
- Agricultural researches along with extension researches should go for frame working, resource monitoring and stewardship functions therewith.
- Baseline surveys and bench marking information are essentials as to what and how the farm energy, fertilizer and agri chemicals, water and bio diversities are exploited, regenerated or eroded to frame up a community level policy with a proper justification for stewardship functions.
- Auditing, costing and Financing after stewardship function should also be matched or complied with ecological deliverables generated by respective stewardship functions at present and in posterity.
- High end statistical packages are to be applied for efficient and high end monitoring of erosion depletion or retrieval of natural resources. 3 dimensional modeling, ARIMA, Time series analysis, sensitivity analysis can help segregating optimizing and customizing the stewardship functions as applicable in a big volume of meta data systems with extractable technical information there from.

Limitations:

- Number of respondents would have been increased but due to the covid 19 situation, the number of respondents has been restricted to 50 numbers. We know the higher would be the number of respondents, the lesser would have been the error and more elaboration of the empirical interpretation has been added.
- More number of variables, relevant and conceptual could have been added to make the study more comprehensive.
- Application of data analytics could have been helpful for future projection of scenario through stimulation.
- Normalization of the data with the application of suitable statistical techniques, could have been effective in making the model more adaptive and effective.

Bibliography

Acharjee S K, Ghosh U, Biswas A and Murah A.K., 2004, A study on the perceived benefits of farm forestry in terms of socio economic and motivational correlates in Coochbehar district of West Bengal, *The Indian Foretser*, Vol.130, No.3, March2004.

Acharya CL, Kapur OC, Dixit SP. Moisture conservation for rainfed wheat production with alternative mulches and conservation tillage in the hills of north-west India. Soil Tillage Res. 1998; 46:153-163.

Adger, W.N. 2006. Vulnerability. *Global Environmental Change* 16:268 261.

Adger, W.N., N.W. Arnell, and E.L. Tompkins. 2005. Successful adaptation to climate change across scales. *Global Environmental Change* 15:57 86.

Adhikary, M.M., K. Pradhan, and S. Mishra. 2013. Analysing and ensuring persuasive extension with respect to preventive innovation for sustainable agriculture. *Journal of Crop and Weed* 9(1):56 68. West Bengal, India: Crop and Weed Science Society.

Affholder FD, Jourdain DD, Quang TP, Tuong M, Morize, Ricome A. Constraints to farmers' adoption of direct-seeding mulch-based cropping systems: A farm scale modeling approach applied to the mountainous slopes of Vietnam. Agricultural Systems. 2010; 103:51- 62.

Agricultural Statistics at a Glance 2014, Government of India, Ministry of Agriculture, Department of Agriculture and Cooperation, Directorate of Economics and Statistics, Oxford University Press.Agriculture census 2010 11 and West Bengal development report 2010. Agricultural Statistics at a Glance, 2014.

Ahnstrom, J., J. Hockert, H.L. Bergea, C.A. Francis, P. Skelton et al. 2009. Farmers and nature conservation: what is known about attitudes, context factors and actions affecting conservation? *Renewable Agriculture and Food Systems* 24(1):38 47. Cambridge, UK: Cambridge University Press.

Allen, and Patricia. 1993. Connecting the Social and the Ecological in Sustainable Agriculture. In *Food for the Future,* ed. Patricia Allen. New York: John Wiley & Sons.

Allison, E.H., B.D. Ratner, B. Asgard, R. Willmann, R. Pomeroy et al. 2012. Rights based fisheries governance: from fishing rights to human rights. *Fish and Fisheries*13(1):14 29. Oxford, UK: WileyBlackwell.

Beer, J., M. Ibrahim, and A. Schlonvoigt. 2000. Timber production in tropical agroforestry systems of Latin America. XXI IUFRO World Congress.

Berkes, F., J. Colding, and C. Folke, editors. 2003. *Navigating Social Ecological Systems: Building Resilience for complexity and Change.* Cambridge University Press,Cambridge.

Berry, W.2005. Renewing husbandary: After mechanization, can modern agriculture reclaim its soul? *Crop Science* 45:1103 1106

Bhattacharyya M, Acharya S k and Chakraborty S K.,2017,Pollinators Unknown: People's perception of native bees in an agrarian district of West Bengal, India, and its implication in conservation. Tropical Conservation Science, Vol 10: 1 14, sage publication.

Birkeland, P.W. 1999. *Soils and Geomorphology.* 3rd Edition. Oxford University Press, New York.

Blann, K. 2006. Habitat in agricultural landscapes: how much is enough? *A state of the science literature review.* West Linn, OR: Defenders of Wildlife.

Bot A, Benites J. Creating drought-resistant soil. In the Importance of Soil Organic Matter: Key to DroughtResistant Soil and Sustained Food Production". FAO soils bulletin 80, FAO Land and Plant Nutrition Management Service, Food and Agriculture Organization of the United Nations Rome, 2005a, 35-40.

Bot A, Benites J. The role of conservation agriculture in organic matter deposition and carbon sequestration. In The Importance of Soil Organic Matter: Key to DroughtResistant Soil and Sustained Food Production, 2005b, 47- 50. FAO Soils Bulletin 80, FAO Land and Plant Nutrition Management Service, Food and Agriculture Organization of the United Nations Rome, 2005.

Bradford JM, Peterson JA. Conservation tillage. In M. E. Summer (Ed.), Handbook of Soil Science (pp. G247eG269). Boca Raton, Florida: CRC Press, 2000.

Brklacich, M.J. 1989. A framework for evaluating the sustainability of food production systems in a changing environment. *Dissertation Abstracts International A, Humanities and Social Sciences* 50(6):p.1764.

Buck, L.E., J.A. Milder, T.A. Gavin, and I. Mukherjee. 2006. Understanding ecoagriculture: a framework for measuring landscape performance. Ecoagriculture discussion papernumber2.

Buck, L.E., T.A. Gavin, D.R. Lee, and N.T. Uphoff. 2004. Ecoagriculture: a review and assessment of its scientific foundations. Ithaca, NY: Cornell University.

Buhaug H, Urdal H. An urbanization bomb? Population growth and social disorder in cities. Global Environ Change. 2013; 23(1):1-10.

Calegari A, Darolt MR, Ferro M. Towards sustainable agriculture with a no-tillage system. Adv. Geo. Ecol. 1998; 31:1205-1209.

Calegari A, Hargrove WL, Rheinheimer DDS, Ralisch R, Tessier D, Tourdonnet S et al. Impact of long-term no of tillage and cropping system management on soil organic carbon in an Oxisol: A Model for Sustainability. Agron. J. 2008; 100(4):1013-1019.

Campbell, A. 2008. Food, energy, water: conflicting insecurities (and the rare win wins offered by soil stewardship). *Journal of Soil and Water Conservation Ankeny.* 63(5):149A 151A. Ankeny, USA: Soil and Water Conservation Society.

Carruthers, D.V. 2001. The politics and ecology of indigenous folk art in Mexico. *Human Organization* 60(4):356 366. Oklahoma City, USA: Society for Applied Anthropology

Cassman, K.G. 1999. Ecological intensification of cereal production systems: Yield potential, soil quality, and precision agriculture. *Proceedings of the National Academy of Sciences* 96:5952 5959.

Castro Filho C, Henklain JC, Vieira MJ, Casão R Jr. Tillage methods and soil and water conservation in southern Brazil. Soil Tillage Res. 1991; 20:271-283.

Chambers, R., and G. Conway. 1991. *Sustainable Rural Livelihoods: Practical Concepts for the 21st century.* IDS, Brighton.

Chapin, F.S., III, M. Hotel, S.R. Carpenter, J. Lubchenco, B. Walker et al. 2006a.

Building resilience and adaptation to manage arctic change. *Ambio* 35:19 202.

Chapin, F.S., III, P.A. Matson, and H.A. Mooney.2002. *Principles of Terrestrial Ecosystem Ecolog.* Springer Verlag, New York.

Chapin,F S.III, Carpenter R.Stephen, Kofinas Gary P., Folke Carl,Abel N, Clark,W C., Olsson, P, Smith D. Mark S ,et. al.,*Ecosystem stewardship: sustainability strategies for a rapidly changing planet.*2009 Elsevier Ltd. All rights reserved. doi:10.1016/j.tree.2009.10.008 Available online 16 November 2009.

Conway,G. 1997.*TheDoubly Green Revolution: Food forAllin the21stCentury.* Cornell University Press, Ithaca.

Cooke, R.J., and J.R. Park. 2001. The use of a decision support tool to assess the financial viability of selected Countryside Stewardship options. *Farm Management*11(3):193 204. Reading, UK: Institute of Agricultural Management, Farm Management Unit, ReadingUniversity.

Daily, and Gretchen, ed. 1997. *Nature's Services: Societal Dependence on Natural Ecosystems.* Washington, D.C: Island Press.

Davies, B.B., and I.D. Hodge. 2007. Exploring environmental perspectives in lowland agriculture: A Q methodology study in East Anglia, UK. *Ecological Economics* 61(2/3):323 333. Oxford, UK: Elsevier.

Deasy, C., J.N. Quinton, M. Silgram, A.P. Bailey, B. Jackson, and C.J. Stevens. 2010. Contributing understanding of mitigation options for phosphorus and sediment to a review of the efficacy of contemporary agricultural stewardship measures. *Agricultural Systems* 103(2):105 109. Oxford, UK: Elsevier Ltd.

De Haan, L. R., Cox, T. S., Van Tassal, D. L., & Glover, J. D.2007 Perennial grains. In Farming with nature: the science and practice of ecoagriculture (eds S. J. Scherr & J. A. McNeely). Washington, DC: IslandPress

Delate, K. 2002. *Using an agroecological approach to farming systems research. HortTechnology*12(3):345 354. Alexandria, USA: American Society for Horticultural Science.

Dia, S., J. Fargione, F.S. Chapin, III, and D. Tilman. 2006. Biodiversity loss threatens human well being. *Plant Library of Science (PLoS)* 4:1300 1305.

Diamond, J. 2005. Collapse: *How Socities Choose or Fail to Succeed.* Viking, New York.

Diaz, S.J., J. Fargione, F.S. Chapin, III, and D. Tilman. 2006. Biodiversity loss threatens human well being. *Plant Library of Science* (PLoS) 4: 1300 1305.

Douglass, and K. Gordon. 1984. Agricultural *Sustainability in a Changing WorldOrder.* Boulder, Colorado: Westview Press.

El Baltagy, A. 1996. West Asia and North Africa: a regional vision. *West Asia and North Africa: a regional vision* 16 pp. Aleppo. Syria: International Center for Agricultural Research in the Dry Areas.

Ellefson, P.V., and M.A. Kilgore. 2010. United States wood based industry: a review of structure and organization. *Staff Paper Series Department of Forest Resources, University of Minnesota* (206):238 pp. St. Paul, USA: Department of Forest Resources, University of Minnesota.

Elmqvist, T., C. Folke, M. Nystrom, G. Peterson, J. Bengtsson, et al. 2003. Response diversity, ecosystem change, and resilience. *Frontiers in Ecology and the Environment* 1:488 494.

Environmental Protection Agency. Emission Facts: Average Carbon Dioxide Emissions Resulting from Gasoline and Diesel Fuel. Available at. US Environmental Protection Agency. 2009. http: / / www.epa.gov / oms / climate / basicinfo.htm (accessed 25 May 2009).

FAO. Soil Conservation and Management for Small Farms. Strategies and Methods of Introduction, Technologies and Equipment. FAO Soils Bulletin No. 77. Rome: FAO, 2000. Farnham, UK: British Crop Protection Council

Fayram, A.H., D.A. Schenborn, J.M. Hennessy, N.A. Nate, P.J. Schmalz. 2009. Exploring the conflict between broad scale and local inland fisheries management: the risks to agency credibility. *Fisheries Bethesda* 34(5):232 236. PB: Bethesda, USA: American FisheriesSociety.

Fischer, J., A.D. Manning, W. Steffen, D.B. Rose, K. Daniell, et al. 2007. Mind the sustainability gap. *Trends in Ecology and Evolution* 22:621 624.

Fischer, J., R. Dyball, I. Fazey, C. Gross, S. Dovers, P.R. Ehrlich, R.J. Brulle, C. Christensen and R.J. Borden. 2012. Human behavior and sustainability. *Frontiers in Ecology and the Environment* 10(3):153 160. PB: Washington, USA: Ecological Society of America. URL:http://www.esajournals.org/doi/full/10.1890/110079

Foley, J.A., R. DeFries, G.P. Asner, C. Barford, G. Bonan, et al. 2005. Global consequences of land use. *Science* 309:570 574.

Folke, C. 2003. Freshwater for resilience: A shift in thinking. *Philisophical Transcation of the Royal Society of London, Series* B 35:2027 2036.

Folke, C. 2006. Resilience: The emergence of a perspective for social – ecological system analysis. *Global Environmental Change* 16:253267.

Folke, C., J. Codling, and F. Brerkes. 2003. Synthesis: Building resilience and adaptive capacity in social ecological system. Pages 352 387 in F. Berkes, J. Clding, and C. Folke, editors. *Navigating Social Ecological Systems: Building Resilience for Complexity and Change.* Cambridge University Press, Cambridge.

Forman, R. T. 1995 Land mosaic: the ecology of landscapes and regions. Cambridge, UK: Cambridge University Press.

Fox, G. 1994. The structure of ownership and prosperity. *Canadian Journal of Agricultural Economics* 42(4):511 524.

Freebairn DM, Boughton WC. Hydrological effects of crop residue management practices. Aust. J Soil Res. 1985; 23: 23-55.

Garrity DP, Akinnifesi FK, Ajayi OC, Weldesemayat SG, Mowo JG, Kalinganire A et al. Evergreen Agriculture: a robust approach to sustainable food security in Africa. Food Security. 2010; 2:197-214.

Gassen DN, Gassen FR. Plantio direto. O caminho do future 207, Aldeia Sul, Passo Fundo, 1996.

Graham, D A; Bonta, M A; Ulloa, R. 2011. Cycad conservation, peasant subsistence, and the military coup in Honduras. *Society and Natural Resources* 24(2):193

Giller KE, Beare MH, Lavelle P, A MN, Izac, Swift MJ Agricultural intensification, soil biodiversity and agroecosystem function. Appl Soil Ecol. 1997; 6:3-16.200. London, UK: Routledge, Taylor & Francis.

Govaerts B, Sayre KD, Deckers J. A minimum data set for soil quality assessment of wheat and maize cropping in the highlands of Mexico. Soil Tillage Res. 2006; 87:163-174.

Govaerts B, Sayre KD, Goudeseune B, De Corte P, Lichter K, Dendooven L et al. Conservation agriculture as a sustainable option for the central Mexican highlands. Soil Tillage Res. 2009b; 103:222-230..

Grigg N.S. 2008. Integrated water resources management: balancing views and improving practice. *Water International.* 33(3):279 292. Carbondale, USA: International Water Resources Association(IWRA).

Gupta R, Gopal Ravi, Jat ML, Jat RK, Sidhu HS, Minhas PS et al. Wheat Productivity in Indo-Gangetic Plains of India During 2010: Terminal Heat Stress and Mitigating Strategies. Conservation agriculture newsletter, Getting agriculture to work for people and the environment, PACA, 1st floor, NASC complex, DPS Marg, Pusa, New Delhi- 110012, India, 2010.

Hancock, D.W. 2012. Grazing schools improve the sustainability of pasture based livestock enterprises. *Journal of Extension*50(5): 5IAW5. West Lafayette, USA: Extension Journal Inc.

Hartmann, M., C.G. Howes, D. VanInsberghe, H. Yu, D. Bachar, et al. 2012. Significant and persistent impact of timber harvesting on soil microbial communities in northern coniferous forests. *ISME Journal* 6(12): 2199 2218. London, UK: Nature PublishingGroup.

Harvey, C.A., J.C. Sanez, and J. Montero. 2005. Conservacion de la biodiversidad en paisajes fragmentados y rurales de Mesoamerica? Que hemos aprendido y que todavı´a necesitamos conocer? Turrialba, Costa Rica: CATIE. Draft.

Hoag, D.L.K and R.B. Boone, and C.M.H. Keske. 2011. The cost for agriculture to coexist with wildlife in Colorado. *Human Dimensions of Wildlife* 16(5):318 329 Philadelphia, USA: Taylor & Francis, Inc.

Hobbs PR, Sayre K, Gupta R. The role of conservation agriculture in sustainable agriculture. Philosophical Transactions Royal Soc. B. 2008; 363:543-555.

Hobbs PR, Govaerts B. How conservation agriculture can contribute to buffering climate change. In M. P. Reynolds (Ed.), Climate Change and Crop Production, 2010, 177- 199. CAB International 2010.

Hobbs PR, Gupta RK. Problems and challenges of no-till farming for the rice-wheat systems of the Indo-Gangetic Plains in South Asia. In R. Lal, P. Hobbs, N. Uphoff, and D. O. Hansen (Eds.), Sustainable Agriculture and the Rice-Wheat System Columbus, Ohio, and New York, USA: Ohio State University and Marcel Dekker, Inc, 2004, 101-119.

Hole, D. G., A.J. Perkins, J.D. Wilson, I.H. Alexander, I. H., Grice, P. V. & Evans, A. D. 2005 Does organic farming benefit biodiversity? *Biol. Conserv.* 122:113–130. (doi:10. 1016/j.biocon.2004.07.018).

Hoogheem, T.J. 1997. Global product stewardship water quality protection.1997 Brighton crop protection conference: weeds. Proceedings of an international conference, Brighton, UK, 17 20 November 1997. 1997; (Volume 3): 10851090.

Hudson BD. Soil organic matter and available water capacity. J Soil Water Conserv. 1994; 49(2):189-194.

Hutsch BW. Tillage and land use effects on methane oxidation rates and their vertical profiles in soil. Bio. Fert. Soils. 1998; 27:284-292

IITA (International Institute of Tropical Agriculture). 2001. Delivering the African agricultural development potential: theIITA vision (Strategic Plan 2001–2010). Ibadan, Nigeria:IITA.

Ikerd, and E. John. 1993. The Need for a Systems Approach to Sustainable Agriculture. *Agriculture, Ecosystems and Environment* 46:147 160.

IPCC (Intergovermental Panek on Climate Change). 2007a. *Climate change 200: The Physical Science Basis, Contribution of Working Group I to the Fourth Assessment Report of the Intergovernmental Panel on Climate Change.* Cambridge University Press, Cambridge.

Jackson, D.L., and L.L. Jackson. 2002. The farm as habitat: reconnecting food systems with ecosystems.Washington, DC: Island Press.

Jackson, J.B.C. 2001. What was natural in the coastal oceans? *Proceedins of the National Academy of Sciences* 9:54115418.

Janssen, M.A., M.L. Schoon, W. Ke, and K. Borner. 2006. Schlorly networks on resilience, vulnerability and adaptation within the human dimensions of global environmental change. *Global Environmental Changes* 16:240 252.

Jat, R.A., K.L. Sahrawat, A.H. Kassam, T. Friedrich. 2014. Conservation agriculture for sustainable and resilient agriculture: global status, prospects and challenges. *Conservation agriculture: global prospects and challenges*1 25. Wallingford, UK:CABI.

Johnson, MG, Levine ER, Kern JS. Soil organic matter: Distribution, genesis, and management to reduce greenhouse gas emissions. Water Air Soil Pollut. 1995; 82:593-615.

Kang, M.S., and S.S. Banga. 2013. Global agriculture and climate change. *Journal of Crop Improvement* 27(6):667 692. Philadelphia, USA: Taylor & Francis.

Kay,P.,A.C.Edwards,andM.Foulger.2009.Areviewoftheefficacyofcontemporary agricultural stewardship measures for ameliorating water pollution problems of key concern to the UK water industry. *Agricultural Systems* 99(2/3):67 75. Oxford, UK: Elsevier.

Kay, P., R. Grayson, M. Phillips, K. Stanley, A. Dodsworth, et al. 2012. The effectiveness of agricultural stewardship for improving water quality at the catchment scale: experiences from an NVZ and ECSFDI watershed. *Journal of Hydrology Amsterdam* 422/423:10 16. Oxford, UK: ElsevierLtd.

Keith M. a,n, Schmidt Olaf b, Barry J. McMahon b. 2016 Soil stewardship as an exus between Ecosystem Services and One Health Aidan, *Elsevier,* A Centre for Ecology & Hydrology, Lancaster Environment Centre, Library Avenue, Bailrigg, Lancaster LA14AP,UK b UCD School of Agriculture & Food Science,

University College Dublin, Belfield, Dublin4, Ireland, Ecosystem Services 17(2016) 40–42, Contents lists available at ScienceDirect

Kleijn, D., A.R. Baquero, Y. Clough, M. Dı´az, J. De Esteban, and F. Ferna´ndez. 2006. Mixed biodiversity benefits of agri environment schemes in five European countries. *Ecol. Lett.* 9:243–254. (doi:10.1111/j.1461 0248.2005.00869.x)

Klein, D.J. and W.J. Sutherland. 2003. How effective are European agri environment schemes in conserving and promoting biodiversity? *Appl. Ecol* 40:947–969. (doi:10.1111/j.1365 2664.2003.00868.x)

Krasny, M.E., and K.G. Tidball. 2012. Civic ecology: a pathway for Earth Stewardship in cities.*Frontiers in Ecology and the Environment*10(5):267 273. Washington, USA: Ecological Society ofAmerica.

Kremen, C., M.N. Williams, and R.W. Thorp. 2002. Thorp crop pollination from native bees at risk from agricultural intensification. *Proc. Natl Acad. Sci.* USA 99:812– 816. (doi:10.1073/pnas.262413599).

Lahiri Dutt, K. 2014. Experiencing and coping with change: women headed farming households in the Eastern Gangetic Plains. *ACIAR Technical Reports Series* (83):66 pp. Canberra, Australia: Australian Centre for International Agricultural Research (ACIAR). URL:http://www.aciar.gov.au/taxonomy/term/15

Lai PoHsin, M.G. Sorice, S.K. Nepal, and Cheng ChiaKuen. 2009. Integrating social marketing into sustainable resource management at Padre Island National Seashore: an attitude based segmentation approach. *Environmental Management* 43(6):985 998. New York, USA: Springer Verlag.

Lambin, E.F. 2007. *The Middle Path: Avoiding Environmental Catastrophe.* University of Chicago Press,Chicago.

Landers, J.N. 2004. Zero tillage challenges and solutions in Brazil. Proceedings VII World Soybean Conference, IV International Soybean Processing and Utilization Conference, III Congresso Brasileiro de Soja Brazilian Soybean Congress, Foz do Iguassu, PR, Brazil, 29 February to 5 March, 2004; 274 281. Londrina, Brazil: Brazilian Agricultural Research Corporation, National Soybean ResearchCenter

Lange, B., and M. Shepheard. 2014. Changing conceptions of rights to water? An eco socio legal perspective. *Journal of Environmental Law* 26(2): 215 242. Oxford, UK: Oxford University Press.

Lu, F., and C. Wirth. 2011. Conservation perceptions, common property, and cultural polarization among the Waorani of Ecuador's Amazon. *Human Organization*70(3):233 243. Oklahoma City, USA: Society for Applied Anthropology.

Mansuri, G., and V. Rao. 2004. Community based (and – driven) development: A critical review. *The World Bank Research Observer* 19(1):139.

Mason, N. 2013. *ODI Working Paper* (363): viii + 48 pp. London, UK: Overseas Development Institute.

Matson, P.A., W.J. Parton, A.G. Power, and M.J. Swift. 1997. Agricultural intensification and ecosystem properties. *Science* 22:504509.

McBratney, A., D.J. Field, and A. Koch. 2014. The dimensions of soil security. *Geoderma* 213:203–213. http://dx.doi.org/10.1016/j.geoderma.2013.08.013

McCarthy, J.J., M.L. Martello, R. Corell, N.E. Selin, S. Fox, et al. 2005. Climate change in the context of multiple stressors and resilience. Pages 945 988 in ACIA, editors. *Arctic Climate Impact Assessment.* Cambridge UniversityPress.

McIntyre, J. 2010. Measuring agricultural stewardship: risks and rewards. The case for the stewardship index for specialty crops. *Journal of Agriculture, Food Systems and Community Development*1(1): 19 22. Ithaca, USA: New Leaf Associates.

McNeely, J.A., and S.J. Scherr. 2003. Ecoagriculture: strategies for feeding the world and conserving wild biodiversity. Washington, DC: IslandPress.

MEA (Millineum Ecosystem Assessment). 2005a. *Ecosystem and Human Being: Current Status and Trends.* Island Press,Washington.

Miller, F.P. 2003. Natural resources management and the second copernican revolution. *Journal of Natural Resources and Life Sciences Education*32:43 51, Madison, USA: American Society ofAgronomy.

Mishra S S, Acharjee S K, Chakraborty S K, 2009, Development of tools for assessing conservation categories of siluroid fishes of fresh water and brackish water wetlands of South West Bengal, India. Environ Biol Fish (2009) 84 : 395 407, @ *Springer.*

Mitchell, D.P.,T.W. Norton, R.I. Grenfell, and P.W. Woodgate. 2007. Barriers to implementation of natural resource management at the landholding level in the intensive agricultural zones of Southern Australia. *Australasian Journal of Environmental Management*14(2):103 110. Melbourne, Australia: Environment Institute of Australia and New Zealand.

Molden, D., R. Tharme, I. Abdullaev, and R. Puskar. 2007. Irrigation. In Farming with nature: the science and practice of ecoagriculture (eds S. J. Scherr & J. A. McNeely).Washington, DC: Island Press.

Molnar, A. 2003. Forest certification and communities: looking forward to the next decade. Washington, DC: Forest Trends. Molnar, A., Scherr, S. J. & Khare, A. 2004, Who conserves theworld's forests? A new assessment of conservation and investment trends. Washington, DC: Forest Trends.

Mooney, Harold and Paul R. Ehrlich. 1997. Ecosystem Services: A Fragmented History. In *Nature's Services: Societal Dependence on Natural Ecosystems*, ed. Gretchen C. Daily, 11 19. Washington, D.C.: Island Press.

Nelson, D.R., W.N. Adger, and K. Brown. 2007. Adaptation to environmental change: contributions of a resilience framework. *Annual Review of Environment and Resources* 32:395 419.

Nierenberg, D. 2005. Happier meals: rethinking the global meat industry. Washington, DC: Worldwatch Institute. Oksman Caldentey, K. M. & Barz, W. H. 2002 Plant biotechnology and transgenic plants. New York, NY: Marcel Dekker.

Odum, E.P. 1989. *Ecology and Our Endangered Life Support Systems.* Sinauer Associates, Sunderland, MA.

Overloop, P.J. van., I.J. Miltenburg, A.J. Clemmens, and R.J. Strand. 2008. Identification of pool characteristics of irrigation canals. Proceedings of the World Environmental and Water Resources Congress, Honolulu, Hawaii, USA, 12 to 16 May 2008, 118. Reston, USA: American Society of Civil Engineers.

Patra M K, Acharjee S K and Chakraborty S K.2005,Conservation categories of siluroid fishes in North East Sundarbans, India, Biodiversity and Conservation (2005) 14: 1863 1876, @ *Springer* 2005.

Peach, W.J. et al. 2001. Countryside stewardship delivers cirl buntings (Emberiza cirlus) in Devon, UK. *Biol. Conserv.* 101:361–373. (doi:10.1016/S0006 3207(01) 000830)

Pearce, D. 2005. Managing environmental wealth for poverty reduction. Poverty and environment partnership MDG7 Initiative economics.

Redford, K.H., and B.D. Richter. 1999. Conservation of biodiversity in a world of use. *Conserv. Biol.* 13:1246–1256. (doi:10.1046/j.1523 1739.1999.97463.x)

Redman, C.L. 1999. *Human Impacts on Ancient Environments*. University of Arizona Press, Tucson.

Regan, J.B. 1995. Natural resource stewardship efficacy of Treflan and Sonalan in conservation tillage. *Down to Earth Midland* 50(2):6 12.

Rhodes, G., P. Henrys, B.C. Thomson, and R.W. Pickup. 2013. Mycobacterium avium subspecies paratuberculosis is widely distributed in British soils and waters: implications for animal and human health. *Environ. Microbiol* 15:2761–2774.

Rotterdam, Netherlands: A.A. Balkema. 2000. Reclaimed land: erosion control, soils and ecology. *Reclaimed land: erosion control, soils and ecology* 385.

Roy B,Rahman Md. Habibur, and Fardusi Most. Jannatul,2013, Status, Diversity and Traditional Uses of Homestead gardens in North Bangladesh : A Means of Sustainable biodiversity conservation, Hindawi Publishing Coorporation, ISRN Biodiversity, Volume 2013, Article ID 124103, 11 pages

Ruark, G., J. Scott, D. Riemenschneider, and T. Volk. 2006. Perennial crops for bio fuels and conservation. Presented at 2006 USA Agricultural Outlook Forum—Prospering in Rural America.

Salve Prachi, How many farmers does India really have? The Hindustan Times ,August 11, 2014.

Sanjines, P.B. 2012. Opportunity analysis of payment for ecosystem services: policy design and implementation for coffee agroforestry systems in Costa Rica. *Tropical Resources: Bulletin of the Yale Tropical Resources Institute*31:56 66. New Haven, USA: Yale School of Forestry and EnvironmentalStudies.

Sara J. Scherr and Jeffrey A. McNeely, Ecoagriculture Partners, *The Royal Society* Washington, DC 2000, USA., World Conservation Union—IUCN, 1196 Gland, Switzerland Phil. Trans. R. Soc. B (2008) 363, 477–494.doi:10.1098/rstb.2007.2165. Published online 25 July 2007

Sayer, J.A., and S. Maginnis. (eds) 2005. Forest in landscapes: ecosystem approaches to sustainability. London, UK: Earthscan.

Schaller, and Neill. 1993. Sustainable Agriculture and the Environment. *Agriculture, Ecosystems and Environment* 46: 89 97.

Scherr Sara J., McNeely Jeffrey A. ,Biodiversity conservation and agricultural sustainability: towards a new paradigm of ecoagriculture' landscapes Ecoagriculture Partners, *Philosophical Transaction of the Royal Society B* (2008) 363, 477–494 doi:10.1098/rstb.2007.2165 Washington, DC 2000, USA 2 World Conservation Union—IUCN, 1196 Gland, Switzerland Published online 25 July 2007

Swaminathan,M. S. (ed.) 1994 Ecotechnology and rural employment: a dialogue. Chennai, India: Macmillan India.

Swift, M.J., A.M.N. Izac, and M. van Noordwijk. 2004. Biodiversity and ecosystem services in agricultural landscapes—are we asking the right questions? *Agr. Ecosyst. Environ.* 104: 113–134. (doi:10.1016/j.agee. 2004.01.013).

Thierfelder C, Wall PC. Investigating conservation agriculture systems in Zambia and Zimbabwe to mitigate future effects of climate change. In Proc. African Crop Science Conf. 2009; 9:303-307.

Thompson, J., T. Hodgkin, K. Atta Krah, D. Jarvis, C. Hoogendoorn, and S. Padulosi. 2007. Biodiversity in agroecosystems. In Farming with nature: the science and practice of ecoagriculture (eds S. J. Scherr & J. A. McNeely). Washington, DC: Island Press.

Ticknor, W.D. 1996. Arkansas forestry: maintaining the momentum of excellence. *SpecialReport Agricultural Experiment Station, Division of Agriculture, University of Arkansas*(175): v + 10 pp. Fayetteville, Arkansas, USA:

Tilman, D. et al. 2001. Forecasting agriculturally driven global environmental change. Science 292:281–284. (doi:10.1126/science.1057544).

Tilman, D., K. Cassman, P. Matson, R. Naylor, and S. Polasky. 2002. Agricultural sustainability and intensive production practices. *Nature* 41:671 677.

Tomich, T.P., M. van Noordwijk, and D.E. Thomas. 2004. On bridging gaps. *Agr. Ecosyst. Environ.* 104:1–3. (doi:10.1016/j.agee.2004.01.002)

Trewavas, A.J. 2001. The population/biodiversity paradox: agriculture efficiency to save wilderness. *Plant Physiol.* 125:174–179. (doi:10.1104/pp.125.1.174)

Turner, B.L., II, R.E. Kasperson, P.A. Matson, J.J. McCarthy, R. W. Corell, et al. 2003. A framework for vulnerability analysis in sustainability science. *Proceedings of the National Academy of Sciences* 100:8074 8079.

Uphoff, N. et al. (eds) 2006 Biological approaches to sustainable soil systems. Boca Raton, FL: CRC Press.

Vitousek, P.M., H.A. Mooney, J. Lubchenco, and J.M. Melillo. 1997. Human domination of Earth's ecosystems. *Science* 277:494499.

Walker, B.H., and D. Salt. 2006. *Resilience Thinking: Sustaining Ecosystems and People in a Changing World.* Island Press, Washington.

Walker, B.H., and J.A. Mayers. 2004. Thresholds in ecological and socio ecological systems: A developing database. *Ecology and Society* 9(2):3.

Walker, B.H., C.S. Holling, S.R. Carpenter, and A.P. Kinzig. 2004. Resilience, adaptability and transformability in social ecological systems. *Ecology and Society* 9(2):5

Walker, B.H., L.H. Gunderson, A.P. Kinzig, C. Folke, S.R. Carpenter, et al. 2006. A handful of heuristics and some propositions for understanding resilience in socio ecological systems. *Ecology and Society* 11(1):13.

Weibull, A.C., O. Ostman, and A. Granqvist. 2003. Species richness in agroecosystems: the effect of landscape, habitat and farm management. *Biodivers. Conserv.* 12:1335–1355.(doi:10.1023/A:1023617117780).

West TO, Post WM. Soil organic carbon sequestration rates by tillage and crop rotation: A Global data analysis. Soil Sci. Soc. Am. J. 2002; 66:1930e1946

Whitten, Stuart, Salzman, James, Shelton, Dave and Wendy Proctor. 2003. Markets for Ecosystem Services: Applying the Concepts. Canberra, Australia: Commonwealth Scientific and Industrial Research Organization(CSIRO).

Williams, J.N., A.D. Hollander, A.T. O'Geen, L.A. Thrupp, R. Hanifin, et al 2011. Assessment of carbon in woody plants and soil across a vineyard woodland landscape. *Carbon Balance and Management* 6(11): (9 November 2011) London, UK: BioMed Central Ltd.

Wittenberg, R., and M.J.W. Cock. 2001. Invasive alien species:a toolkit of best prevention and management practices. Wallingford, UK: CAB International.

Wojtkowski, P.A. 2004. Landscape agroecology. New York, NY: Food Products Press.

WRI (World Resources Institute). 2005. *World Resources 2005: The Wealth of the Poor–Managing Ecosystemsto Fight Poverty* World Resources Institute, Washington.

Young, O.R., F. Berkhout, G.C. Gallopin, M.A. Janssen, E. Ostrom, and S. van der Leeuw. 2006. The globalization of socio ecological systems: An agenda for scientific research. *Global Environmental Change* 16:304 316.

Index

L

M

O

P

R

S

W

9 789359 191065

Printed at Repro India Ltd.